Advance Praise for
Power from the Wind

Dan Chiras is spot on with *Power from the Wind*. This authoritative and smartly researched book is written from first-hand experience and provides an up-to-date lens by which to examine small wind turbine options— from the turbines and towers to the grid-tied inverters or batteries for an off-grid system. He distills it all so that you can make the best decision when evaluating the prospects of your own wind turbine system.

— John D. Ivanko and Lisa Kivirist, co-authors of *ECOpreneuring* and
owners of Inn Serendipity, completely powered by the wind and sun.

Dan pours his heart and soul into his books and this one is no exception! His attention to detail shines through as he covers all aspects of small wind and presents it in an easy to understand way. *Power from the Wind* is a great resource for beginners and seasoned wind veterans alike and it's required reading for all my wind workshop students.

— Roy Butler, Four Winds Renewable Energy, LLC,
Small wind curriculum developer and instructor

Wind energy is a hot topic today as we seek energy alternatives to fossil fuel. But there are a lot of misconceptions about its efficacy and appropriateness, varying by site. In *Power from the Wind*, Dan Chiras provides everything you need to know to determine whether wind is an option for you. From assessing your site, to covering the various system options, through the additional elements necessary to create a complete system, and the potential costs and benefits, Dan explores this topic with his usual rigor and depth of research, while writing it in a very approachable and understandable manner. I cannot more highly recommend this book. If you are wondering whether wind power may be right for you, you could not buy a better book.

— James Plagmann, Architect + LEED AP,
HumaNature Architecture, LLC

When you need practical advice from a warm, smart and informed human being, Dan Chiras is the one to turn to.

—Bruce King, PE Director, Ecological Building Network; and author,
Buildings of Earth and Straw and *Making Better Concrete*

Dan Chiras has done as much as anyone in America to promote and popularize the use of renewable energy.

—StephenMorris, publisher and editor, *Green Living: A Practical Journal for the Environment*;
and editor, *The New Village Green: Living Light, Living Local, Living Large*

Dan Chiras is one of the most authoritative writers in the field of renewable energy. His multiple books create a comprehensive library for homeowners looking to live a lifestyle in harmony with their values.

—David Johnston, author, *What'sWorking: Visionary Solutions for Green Building*
and, *Green Remodeling: Changing theWorld One Room at a Time*

Power
from the
Wind

Power
from the
Wind

Achieving Energy Independence
A Practical Guide
to Small-Scale Energy Production

REVISED 2ND EDITION

DAN CHIRAS

Artwork by Anil Rao, Ph.D. and Forrest Chiras

new society
PUBLISHERS

Cover design by Diane McIntosh.
Cover image: Bergey Excel, courtesy Bergey Windpower Co.
p. 1 © nasik; p. 1 © SidorArt; p. 11 © basketman23/Adobe Stock.

Funded by the Government of Canada   Financé par le gouvernement du Canada | Canada

Printed in Canada by Friesens.
First printing May, 2017.

Inquiries regarding requests to reprint all or part of *Power from the Wind* should be addressed to New Society Publishers at the address below.

To order directly from the publishers, please call toll-free (North America) 1-800-567-6772, or order online at www.newsociety.com

Any other inquiries can be directed by mail to:

New Society Publishers
P.O. Box 189, Gabriola Island, BC V0R 1X0, Canada
(250) 247-9737

LIBRARY AND ARCHIVES CANADA CATALOGUING IN PUBLICATION

Chiras, Daniel D., author

Power from the wind : achieving energy independence : a practical guide to small-scale energy production / Dan Chiras.—Revised 2nd edition.

Includes bibliographical references and index.
Issued in print and electronic formats.
ISBN 978-0-86571-831-9 (softcover).—ISBN 978-1-55092-626-2 (ebook)

1. Wind turbines. 2. Wind power. 3. Electric power production. I. Title.

TJ828.C45 2017   621.31'2136   C2016-908243-1
                                         C2016-908244-X

New Society Publishers' mission is to publish books that contribute in fundamental ways to building an ecologically sustainable and just society, and to do so with the least possible impact on the environment, in a manner that models this vision.

I dedicate this book to Judith Plant and the late Chris Plant,
my publishers and friends for many years.
Chris and Judith are visionaries—passionate about making change,
and they have contributed enormously through the many books they have published,
a tradition being aptly carried on by the new employee-owners of New Society,
a crew I love dearly and hold in extraordinarily high esteem.

Contents

Preface

My work on this book started in the summer of 2006. I'd just published *The Homeowner's Guide to Renewable Energy*, a book that helps readers understand their options for tapping into renewable energy such as wind energy, solar electricity, and passive solar heating/cooling technologies.

Emails and phone calls started arriving almost immediately, with technical questions about wind. Readers assumed I was an expert in every renewable energy technology I discussed in the book, when, in fact, my area of greatest expertise at the time was in passive solar heating and cooling. I'd published a book a few years earlier on that topic entitled *The Solar Home: Passive Heating and Cooling*. Although I'd installed my own wind system, had a solar hot water collector installed in a previous home, and had lived off-grid on solar electric in my super-efficient passive solar home for many years, my knowledge of wind and renewable energy systems other than passive solar left a little to be desired.

In the summer of 2006, a few months after *The Homeowner's Guide to Renewable Energy* was published, I was visiting a client in northern Michigan, helping her design a super-efficient passive solar facility for a small organic farm and ecological learning center. She asked if I'd help her design a wind energy system to produce electricity to pump water to irrigate her crops. I said "yes," reflexively. What could be so hard about that?

On the flight home, however, the magnitude of my commitment hit home. So, shortly after I got home, I began to read everything I could put my hands on about small wind to expand my knowledge. My interest in small wind energy systems became a near obsession. This amazing technology, with all its technical intricacies, enthralled me. I wanted to know more.

As my knowledge grew, I felt the urge to put it all down on paper, to write, ah yes, *another* book—this one would be for those who'd like to learn about wind but don't want to wade through the handful of technical books and dozens upon dozens of articles on the subject. As my knowledge grew, however, so did my awareness that this was a subject that required an extremely high level of expertise. I quickly came to realized that I needed a really smart, experienced, and knowledgeable expert in small wind energy to help me—to provide advice and guidance, correct inadvertent mistakes, and provide additional information. While attending the Midwest Renewable Energy Association's annual energy fair, I asked Mick Sagrillo, a guy who fit the bill precisely, if he'd help. Mick's the guru of small wind. He'd been in the small wind business longer than just about anyone. Mick said "yes." Despite his hectic schedule, he assisted me with the first edition, for which I am eternally grateful.

A year or so later, I approached New Society Publishers with the idea of publishing a series of books on renewable energy technologies for the home and office. They liked the idea, and we were off and running.

After bringing Mick on, I signed up for numerous workshops on wind energy through various entities, including the Midwest Renewable Energy Association and the American Solar Energy Society. I attended many workshops on small wind energy that covered a wide variety of topics, including wind turbine design, wind site assessment, tower design and construction, and the installation of wind turbines. In late 2007, I became a certified wind site assessor—at the time only one of 12 in the nation.

That year, I also recruited another wind expert, *Home Power*'s Ian Woofenden. Like Mick, Ian teaches installation workshops and writes articles for *Home Power* magazine on small wind energy. Ian supplies a good portion of his own electricity with wind and solar; he's been living off grid for many years. He also helped me prepare the first edition of this book. He proofed chapters, answered questions, and offered new insights.

In the summer of 2007, Jim Green, the National Renewable Energy Laboratory's small-wind expert asked if he could help, too, after I attended a workshop on small wind energy that he co-taught with Abundant Renewable Energy's Robert Preus at Solar 2007 in Cleveland. Jim volunteered to read the manuscript and provide his insights, to help ensure that the technical accuracy of the book.

A few months later, I recruited another expert, Robert Aram, an electrical engineer, who I met and worked with in several wind energy workshops. Bob has a vast knowledge and an extraordinary ability to explain things clearly. He said that he'd be happy to read the book to help ensure technical accuracy. Bob impressed me with his precision and his no-nonsense insistence on correct math, physics, and engineering terms. I hired him to give me a hand in such areas, to be sure that my book, while written for the non-engineer, was correct in every aspect.

Mick, Ian, Jim, and Bob offered extremely valuable comments that helped me produce what I felt was the most accurate, up-to-date, and readable resource on small wind energy. Without them, the task would have been impossible!

I am deeply indebted to all of them for their tactful yet honest comments and dedication to accuracy and am extremely honored to have worked with such amazingly dedicated and knowledgeable people. A world of thanks to them and to all the others whose work I relied on when researching and writing the first edition of this book.

By 2016, it became clear that I needed to revise the book to reflect the many changes I'd witnessed in the industry. I had also gained considerable experience installing and even making my own small wind turbines. And I had gained experience with everyday living on wind energy; my partner Linda and I had powered our home in Missouri with a small wind turbine for quite a few years by then. I wanted to share some of these experiences with my readers.

This new edition, which you hold in your hands, is a product of all the work I've done, the experience I've gained, and the generous assistance of my early book consultants. I'd like to extend a world of thanks to them and to my friends and colleagues at New Society Publishers, who, over the years, have been an absolute delight to work for. Thanks for agreeing to take on the revision of this book and for believing in me, and for supporting my many trips to *Mother Earth News* Fairs, where I often give talks on solar and small wind. And many thanks for their unwavering dedication to creating a new, sane, just, and sustainable society. Many thanks to my dear friends Ingrid and Sue for all they've done to make this a better book. As always, I'm very grateful to Greg Green, who has worked with me on interior design for a dozen or so of my books, and to John McKercher, who does a smashing job of translating design into layout. Thanks, too, to my copyeditor, Linda Glass, who has also worked with me on numerous books. Her attention to detail and suggestions have helped make this a much better book. I'd also like to thank, Dr. Anil Rao, a professor of biology, who illustrated the first edition of this book for me; and thanks to my son Forrest who added numerous new illustrations in this book. Both have been a pleasure to work with and a valuable component of this book's success.

A world of thanks goes to my partner Linda, who has stood by me through thick and thin for over 20 years. She deserves a medal of honor; life with a writer can be a challenge. And many, many thanks to the rest of my (very patient) family.

—Dan Chiras, Ph.D.
The Evergreen Institute
Gerald, Missouri
November, 2016

CHAPTER 1

Introduction to Small-scale Wind Energy

Humans have harvested energy from the wind for centuries. Prior to the advent of steam-powered ships, for example, Phoenicians, Europeans, and others relied on the wind to propel magnificent sailing vessels across a largely uncharted planet. Ships then became an important mode of transport for raw materials and finished products to and from Europe.

Our predecessors also used wind to assist in food production and to manufacture goods. The windmills of Europe, for example, which were in place 800 to 900 years ago, were used to grind grain into flour to feed Europe's masses. The Dutch used wind to pump water from coastal wetlands, so they could be converted to farmland to grow food.

Wind energy has a long history in North America, too, stretching into the late 1800s. During this period, windmills on tens of thousands of farms in the Great Plains of North America pumped water for livestock, gardens, and humans. Without them, many farmers would not have been able to provide sufficient water for their cattle and sheep.

In the 1890s, more than 100 manufacturers were producing water-pumping windmills in the United States, notes small-wind expert Jim Green of the National Renewable Energy Laboratory (NREL). Both wind-electric generators and water-pumping windmills were extremely popular among farmers and ranchers. According to the NREL, over 8 million mechanical windmills (water pumpers) were installed in rural America, beginning in the 1860s (Figure 1.1). Many of these water-pumping windmills have been restored and are

> ### Windmill vs. Wind Turbine
>
> A windmill is a piece of equipment that drives a mechanical device such as a water pump or a grindstone. A wind turbine drives an electrical generator.

still operating today, providing many more years of reliable service with minimal maintenance.

Although history books make little mention of it, in the 1920s through the early 1950s many Plains farmers also installed small wind turbines to generate electricity. These electricity-generating wind turbines made life on the Great Plains more bearable. Homegrown electricity was used to power lights and a few modern conveniences, among them electric toasters, washing machines, and radios—all ordered from the Sears catalog. The radio was highly coveted as a way of keeping in touch with the world; Sears customers who purchased a radio were given a discount on a wind generator.

Wind energy was not only vital to farmers, it was extremely important to railroads. Windmills were often used to fill water tanks along tracks to supply the steam engines of early locomotives.

Unfortunately, the use of water-pumping windmills and wind-powered electric generators began to decline in the United States in the late 1930s. The demise of these technologies was due in large part to America's ambitious Rural Electrification Program.

This program, which began in 1937, was designed to provide electricity to rural America. As electric service became available, wind-electric generators were mothballed. In fact, local power companies required farmers to dismantle their wind generators as a condition to their provision of service via the ever-growing electrical grid. The electrical grid, typically simply referred to as *the grid*, is the extensive network of electrical transmission lines that crisscross our nation, delivering electricity generated by centralized power plants to cities, towns, and rural customers. Initially, a key advantage of the grid was its ability to provide virtually unlimited amounts of electricity to those who had the wherewithal to pay for it. The grid also made it possible to power large motors, something that wind/battery systems were unable to do.

FIGURE 1.1. Water-pumping windmills like this one, photographed by Dan on a commercial wind farm in southeastern Colorado, were once common through the West and Midwest. The technology is so good that it hasn't changed in 100 years. Credit: Dan Chiras.

Although farmers' lives improved as a result of rural electrification, once-profitable manufacturers of wind-electric generators were driven out of business by the early 1950s. In the mid 1970s, however, wind energy made a resurgence as a result of intense interest in energy self-sufficiency in the United States and elsewhere. This new-found interest in self-reliance was stimulated principally by back-to-back oil crises in the 1970s that resulted in skyrocketing oil prices and a period of crippling inflation in the United States. Generous federal incentives for small wind turbines (a 40 percent US federal tax credit), equally charitable incentives from some state governments, and changes in US law that required utilities to buy excess electricity from small renewable energy generators helped spark the comeback. From 1978 to 1985, 4,500 small utility-connected residential wind turbines were installed, according to Mike Bergey, whose company Bergey WindPower manufactures small wind turbines. In addition, approximately 1,000 wind turbines were installed in remote locations not connected to the electrical grid.

In short order, however, wind energy's resurgence died, falling victim to economic forces beyond its control. Energy-efficiency measures in the United States and new, more reliable sources of oil from Great Britain, Russia, and other countries, drove the price of energy downward. These factors, combined with the

FIGURE 1.2. Mick Sagrillo, seen here perched on a lattice tower, has been in the small wind industry since 1981. He has served as a mentor and advisor to me on this project, for which I am most grateful. Credit: Dan Chiras.

end of federal and state renewable energy tax incentives and a dramatic shift in the political climate away from renewable energy in the early 1980s, resulted in a precipitous decline in America's concern for energy independence. As a result of these changes, most of the fledgling wind manufacturers went out of business. In fact, six years after the end of the tax credits, virtually all of the 80 or so wind generator companies doing business in the United States disappeared, according to Mick Sagrillo, a small-wind-energy expert who served as a technical advisor for the first edition of this book.

In the 1990s, commercial and residential wind energy made another comeback. This rise in the popularity of wind and other renewable energy resources was spurred by a concern over rising energy prices, declining fossil fuel resources, and growing evidence of global climate change and its many social, economic, and environmental impacts.

Much to the delight of renewable energy advocates, large commercially operated wind farms are popping up on land and in the sea in numerous countries, most notably the United States, Germany, Spain, Denmark, and, more recently, China. Commercial wind farms generate huge amounts of electricity and have significantly changed the way we meet our energy needs. In fact, wind-generated electricity is currently the fastest growing source of energy in the world (Figure 1.3). The United States is a leader in wind energy production. According to the American Wind Energy Association (AWEA), in 2015 there were nearly 1,000 utility-scale wind projects—large wind farms—and over 48,800 large commercial wind turbines installed in 40 US states plus Puerto Rico and Guam. Moreover, there were "more than 500 wind manufacturing facilities spread across 43 states." The United States now generates over 13 percent of its electricity from renewables with over one-third of that electricity coming from large commercial wind farms. Several states lead the way.

FIGURE 1.3. This graph shows the installed global capacity (in megawatts) of commercial wind turbines. Notice the dramatic increase since 2000. Credit: American Wind Energy Association.

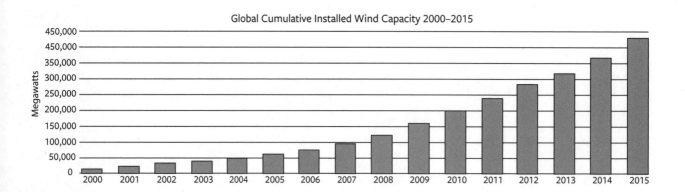

Global Cumulative Installed Wind Capacity 2000–2015

Rated Power and Capacity

Wind turbines are commonly described in terms of rated power, also known as rated output or rated capacity. Rated power is the instantaneous output of the turbine (measured in watts) at a certain (rated) wind speed and at a standard temperature and altitude.

To understand what this means, let's first explore *watts*. Most of us are familiar with the electrical term watts because we've purchased light bulbs and a host of other devices like hair dryers and microwaves that are rated by their wattage.

Watts is a measure of instantaneous power consumption by an electrical device, known as a *load*. The more watts a device consumes, the more energy it consumes, and the more it costs you to operate it.

Watts can also be used to describe the output of electricity-generating devices such as wind turbines, solar modules, and conventional power plants. A wind turbine, for instance, might produce 3,000 watts under moderate winds, but 10,000 watts under winds of 25 miles per hour (about 11 meters per second).

Small wind turbines, the subject of this book, have a rated power of 1,000 to 100,000 watts. Because 1,000 watts is one kilowatt (kW), small wind falls in the range of 1 to 100 kilowatts. Large wind turbines include all of those turbines over 100 kilowatts. Most larger turbines in operation today are 1 megawatt and larger turbines—typically around 1.5 megawatts. A megawatt is a million watts or 1,000 kilowatts. (By the way, a 1.5 megawatt wind turbine can produce enough electricity for 300 to 900 homes, depending on the average wind speed at the site and homeowner energy consumption.)

While rated power has been used to categorize wind turbines for many years, it is one of the least useful and most misleading of all parameters by which one should judge a wind generator's performance. That's because for many years manufacturers rated their turbines at different wind speeds. One manufacturer might rate its wind turbine at 27 miles per hour (12 m/s); another might rate his turbine at 25 miles per hour (11.2 m/s). This made it extremely difficult to compare one turbine to another.

Thanks to the efforts of numerous small wind energy advocates, the industry has recently begun to standardize wind turbine rated output—measuring them all at 11 meters per second (24.6 miles per hour). This makes it much easier to compare one wind turbine to the next.

As you study wind energy and other energy systems, you'll commonly hear experts talk about the "capacity" of a wind turbine. A 20-kW wind turbine, for instance, will produce 20,000 watts, but only at its rated wind speed of 11 m/s. (However, winds at that speed aren't a very common occurrence.) You will also hear talk about things like 300-megawatt wind farms. The capacity of a wind turbine is calculated by multiplying the rated output of each wind turbine by the number of turbines. Two hundred 1.5 megawatt commercial wind turbines would produce a 300-megawatt wind farm. Bear in mind, however, that a wind farm of this size would not produce 300 million watts of electricity at all times. It would only do so when all the wind turbines were operating at their rated wind speed—usually in the mid 20 mile-per-hour range.

That said, capacity is still a handy number to know. For example, many people ask me: "What size wind turbine will I need to power my home?" I tell them that most homes can be powered with wind turbines in the 10 to 20 kW range. That's the approximate turbine size a homeowner needs. Of course, it's more complicated than that, but this gives you a good idea of the range you might require. (More on this topic in Chapter 4.)

Iowa, for instance, in 2015, generated over 30 percent of its electricity from large commercial wind farms. Kansas and South Dakota generated more than 20 percent of their electricity from wind.

Although commercial wind farms are responsible for nearly all of the growth in the wind industry, smaller residential-scale wind turbines have also been popping up in rural parts of America and other countries, supplying electricity to homes, small businesses, farms, ranches, and schools (Figure 1.4). Even a few large businesses have installed small wind turbines (under 100 kilowatts) to power their facilities. Most of the small-scale wind turbines in the United States "feed" the excess electricity they produce back onto the electrical grid. However, most of the small wind turbines produced worldwide are manufactured for off-grid applications. In Canada, for instance, many small wind energy systems power remote off-grid homes in the isolated northern reaches of the country.

Ranchers and farmers sometimes use wind turbines to supply power to electric fences, stock watering tanks, and remote lighting—that is, small dedicated loads to which it is not cost effective to run a power line. (A load is any device that consumes electricity.) I've seen small wind turbines used to power park facilities in remote locations in Alaska. Many sailboats are equipped with very small wind turbines (under 1,000 watts—typically referred to as *microturbines*) to power lights, fans, and refrigerators (Figure 1.5).

Wind energy is being tapped to power remote villages in less developed countries, where the cost of stringing power lines from centralized power plants is prohibitive. Wind energy has even found a home in remote sites in some developed countries. In France, for instance, the government paid for the installation of wind turbines and solar-electric systems on farms at the base of the Pyrenees—rather than running electric lines to these remote sites. Reportedly, tens of thousands of nomads in Mongolia own tiny wind generators that provide electricity to their yurts to power lights and, get this, small televisions. When a family moves every few weeks in search of new pasture for their livestock, their small, durable wind generators are packed

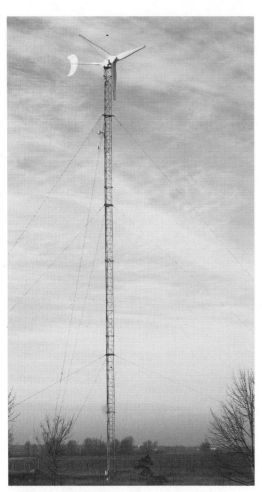

FIGURE 1.4. Small Wind Turbine on Tower. This ARE442 wind turbine installed at Mick's house during a workshop is mounted on a guyed lattice tower. Maintenance is performed by climbing the tower. The ARE442 is now manufactured and sold by a company called Xzeres. The turbine is called the Xzeres 442SR. Credit: Dan Chiras.

FIGURE 1.5. Wind Turbine on Sailboats. (*a*) Microturbines, such as the one shown here by Marlec, are frequently used on sailboats to charge batteries that supply electricity for loads such as radios, lights, televisions, and refrigerators. (*b*) Microturbines for marine use are designed to withstand the harsh environment. This one is made by Aerogen. Credit: (*a*) Marlec and (*b*) Dan Chiras.

up and transported on the backs of pack animals. For newlyweds in this part of China, a small wind turbine is a highly coveted gift.

Wind clearly has a long history of service to humankind, and it is on the rise. Proponents say it could become a major source of electricity in years to come.

World Wind Energy Resources

Although wind energy's popularity is at an all-time high and continues to grow yearly, what is its potential? Can wind become a major source of energy in the future?

Wind is a ubiquitous resource. Although not evenly distributed throughout the world, significant resources are found on every continent. Globally, wind resources are phenomenal. Tapping into the world's windiest locations could theoretically provide 13 times more electricity than what is currently produced worldwide, according to the Worldwatch Institute, a Washington, DC-based nonprofit organization that's played a huge role in creating a sustainable future.

In North America, wind is abundant much of the year in the Great Plains and in many northern states. It is also a year-round source of energy along the Pacific and Atlantic Oceans and the shores of the Great Lakes. Tapping into a couple of the windiest locations in the United States—for example, the states of North and South Dakota—could produce enough electricity to supply *all* of the nation's electrical needs. Proponents of wind energy, like the Worldwatch Institute, estimate that wind energy could provide 20 to 30 percent of the electricity consumed in many countries. Others believe that wind could provide an even larger percentage.

Could wind provide 100 percent of the world's electrical energy needs?

Yes, it could, theoretically...*if* we had more efficient ways of storing and then transferring electricity onto an electrical grid.

Will it?

Probably not.

Other sources of renewable and nonrenewable energy will also come into play. As Ian Woofenden, wind energy expert, author, and technical advisor to the first edition of this book points out: "The 'Can wind do it all?' question is a bit of a red herring. Wind is one piece in the puzzle; nothing is the whole answer." In the future, electrical production will no doubt be provided by a number of renewable energy sources. Wind will very likely play a huge role in many parts of the world, but large commercial solar-electric facilities and solar-electric systems on homes and commercial buildings will also produce a significant amount of electricity.

The potential of the Sun, like that of the wind, is nothing short of phenomenal. It's estimated, for instance, that the sunlight striking an area the size of the state of

Connecticut could meet all of the United States' (inefficient and wasteful) electrical demand. Although no one is proposing the construction of such large solar-electric arrays, solar-electric modules on homes, office buildings, schools, and commercial solar-electric facilities in the best locations could provide an enormous amount of electricity, greatly supplementing wind energy production (Figure 1.6).

Geothermal and biomass resources could contribute their share as well. Biomass resources refer to plant matter such as wood chips that can be burned directly to produce heat to generate steam to make electricity. Plant matter such as corn can also be converted into gaseous or liquid fuels that can be burned to create electricity. Animal wastes can also be used to generate methane, the main component of natural gas.

Hydropower will continue to do its part in the future, and lest we forget, conventional fuels such as oil, natural gas, coal (burned as cleanly as possible), and nuclear energy will also be part of the mix for many years to come. I suspect they will gradually transition into being sources of backup power, supplementing renewables.

Renewable energy, including wind, is here to stay and will likely contribute even more energy to power our future. It has to for the simple reason that fossil fuels are limited. Oil could be economically depleted within 30 to 50 years. Production

FIGURE 1.6. Solar Array. In a renewable energy economy, large-scale solar-electric installations, like the one at Nellis Air Force Base in Nevada shown here, will supplement electricity produced by other renewable resources, including wind, hydropower, and biomass, as well as conventional fuel sources. Credit: David Amster.

rates worldwide are on the decline now. Natural gas production could also peak in the not-too-distant future. The Sun, however, which powers solar energy systems and creates winds that can be tapped by wind turbines, is going to be around for at least another five billion years.

The Pros and Cons of Wind Energy

Wind is a seemingly ideal fuel source that could ease many of the world's most pressing problems. Like all energy sources, wind power has its advantages and disadvantages. Let's look at its downsides first.

Disadvantages of Wind Energy

As you read the downsides of wind energy, you'll discover that many of them pertain to large commercial wind projects. These concerns, in turn, often trickle down unfairly to small wind—the kind of system you are no doubt contemplating. You'll also see that, while there are legitimate problems with wind energy, many are only *perceived* problems—"problems" that result from misconceptions, ignorance, and, frequently, outright deception on the part of opponents. I'll be sure to point these out as we proceed.

Variability and Reliability of the Wind

Perhaps the most significant "problem" with small- and large-scale wind energy is that the wind does not blow 100 percent of the time in most locations. Like solar energy, wind is a variable resource. A wind turbine may operate for four days in a row, then sit idle for the next two days. In most locations, winds are typically strongest in the fall, winter, and early spring, but die down during the summer months.

Wind even varies during the course of a day. Winds may blow in the morning, then die down for a few hours, only to pick up later in the afternoon and blow throughout the night.

Even though wind is a variable resource, it is not unreliable. Just like solar energy, you can count on a certain amount of wind each year. With smart planning and careful design, you can design a wind system to meet some or all of your electrical needs.

Wind is also predictable. With advanced weather forecasting, it's easy to know when it will be windy and when it won't. This allows utilities to integrate wind into their existing system.

Moreover, on a commercial level, wind turbines are most often installed in the windiest locations—places where the winds blow 65 to 85 percent of the time—for

example, along coastlines or in the Plains states in the United States. Residential wind turbines provide the most reliable and economical power when installed in similar locations. Keep that in mind when considering this renewable energy option.

Wind's variable nature can be managed to the benefit of off-grid system owners with the installation of batteries to store surplus electricity. The stored electricity can power a home or office when the winds fail to blow—or when demand exceeds the output of the turbine.

Surplus wind-generated electricity produced by grid-tied systems can also be "stored" on the electrical grid. That is, when a wind-electric system is producing more power than a home or business is using, the excess can be backfed onto the grid. When a wind turbine is not producing at all or is producing less than is required, electricity is drawn from the grid.

On a commercial level, wind energy surpluses generated in one region of a country can be used to offset shortages in another. For example, in Colorado, wind farms in the northern part of the state may be active while wind farms in the southern part of the state are not. Electricity from the former as well as wind farms in neighboring Wyoming could ensure a steady supply of renewable energy to residents. Wind can be integrated into an electric grid supplied by large-scale solar facilities, too. These concepts are illustrated in Figure 1.7.

Wind's variable nature can also be offset by coupling small wind systems with other renewable energy sources, for example, solar-electric systems or micro hydro systems. These are referred to as *hybrid systems*. Solar-electric systems or photovoltaic (PV) systems generate electricity when sunlight strikes solar cells in solar modules. Micro hydro systems tap the energy of flowing water in nearby streams or rivers. They convert this energy into electricity. Hybrid systems can be designed to provide a reliable year-round supply of electricity. As you shall see in Chapter 3, residential wind-generated electricity can also be supplemented by small gas or diesel generators.

> As a power source, wind energy is less predictable than solar energy on a day-to-day basis, but it is also typically available for more hours in a given day.
>
> Mike Bergey,
> Bergey WindPower

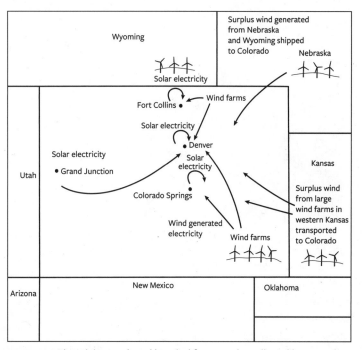

FIGURE 1.7. Electricity produced by wind farms and small- and large-scale solar operations can be shipped from areas of surplus to areas in need of energy via the electrical grid, helping create a reliable source of energy throughout a country. Credit: Forrest Chiras.

The topic of wind's presumed unreliability is even raised when comparing different renewable energy systems—for example, when wind is compared to solar electricity. Some individuals mistakenly view solar electricity as a more reliable resource than wind. However, the wind is much more predictable than you'd think. To understand what I mean by this, let's look at *capacity factor*, a measurement used to compare different electrical generating technologies.

Capacity factor is the ratio of the output of a power plant over some period to what its output would have been had it operated at its rated power for the same period. For example, let's suppose you live in an area with an average of five peak hours of sunlight for PV production per day (for example, Kansas City). In this location, the capacity factor for PV would be 5 hours per day divided by 24 hours per day, or about 21 percent.

In the lower 48 states, the capacity factor for most PV systems ranges from 8 to 25 percent. According to wind expert Mick Sagrillo, the capacity factor for small wind systems ranges from 10 to 28 percent. So, PV and wind systems are fairly similar.

The capacity factor for wind is higher because wind turbines can work day or night. They can operate on sunny or cloudy days—so long as the wind is blowing. What is more, because wind and sunlight are often available at different times, the two technologies complement each other extremely well. Hybrid systems increase the electrical energy produced at a site and ensure steadier supply of electricity.

Site Specific

Yet another criticism of wind—often lodged by solar proponents—is that wind energy is more site specific—or restricted—than solar energy.

To understand what this means, let me point out that there are good (sunny) solar areas and good (windy) wind areas. In a good solar region, most people with a good southern exposure can access the same amount of sun. In a windy area, however, hills and valleys or stands of trees can dramatically reduce the amount of wind that blows across a piece of property. Therefore, even if you live in an area with sufficient winds, you may be unable to tap into the wind's generous supply of energy because of topography or vegetation like tall stands of trees. That's what critics mean when they say that wind energy is more site specific.

That said, I'd be remiss if I did not point out that solar resources also vary. If you live in a sunny region but your home is located in a forest,

FIGURE 1.8. Ian Woodfenden and His ARE110 (no longer manufactured). Perched on top of this 168-foot tower is wind energy expert, author, and workshop teacher Ian Woofenden, who served as a primary technical advisor on the first edition this book. This extremely tall tower raises the turbine well above the trees that carpet the island where Ian lives, allowing access to the wind and permitting excellent performance. Credit: Shawn Schreiner.

you'll receive less solar energy than a nearby neighbor whose home is in an open field. Note, though, that homeowners can access the wind at less-than-optimum sites by installing turbines on tall towers. Ian Woofenden, for example, installed a turbine at his home which is nestled in a densely forested island in the Pacific Northwest. He made it work by installing the turbine on a 168-foot tower—well above the tops of the trees. Tall towers help us overcome topographical and other barriers. They can also be used to augment or "magnify" the wind. That is, an individual can harvest more wind energy by increasing tower height. As Mick Sagrillo points out in his wind energy workshops, you can't make a location sunnier, but by increasing tower height you can move a turbine into smoother, higher velocity winds to boost its output.

Bird and Bat Mortality

Another perceived problem that frequently arises in debates over wind energy—particularly large wind energy—is bird and bat mortality. This issue has been blown way out of proportion. Although a bird may occasionally perish in the spinning blades of a residential wind turbine, this is an extremely rare occurrence. Renewable energy expert Ian Woofenden is aware of only one instance of a bird kill, when a hawk flew into a small wind turbine. "Because of their relatively smaller blades and short tower heights, home-sized wind turbines are considered too small and too dispersed to present a threat to birds," notes Mick Sagrillo in his article, "Wind Turbines and Birds" published by Focus on Energy, Wisconsin's energy efficiency and renewable energy program.

The only documented bird mortality of any significance occurs at large commercial-scale wind turbines—but even then, the number of deaths is extremely small. In my view, the argument that wind energy development should be halted because of bird kills is ill-informed; in fact, it is often a ploy used by individuals and organizations that oppose wind energy development. In editorials and public hearings, opponents often use inflammatory language to make their case, calling wind turbines "bird blenders" or "eagle killers." Outlandish numbers of deaths are often attributed to them.

If citizens and governments were serious about bird kills, we'd ban the truly lethal forces discussed in the accompanying textbox: domestic cats, utility transmission towers, cars, pesticides, and windows (Figure 1.9). We'd even prohibit farming, which destroys bird habitat and poisons birds with pesticides.

FIGURE 1.9. Cats are the leading cause of bird deaths. Our kitty is perched on top of a bluebird nest box at my home in Evergreen, Colorado. Credit: Linda Stuart.

Bird Kills from Commercial Wind Farms: Fact or Fiction?

While commercial wind turbines do kill a small number of birds, scientific studies show that the problem has been grossly exaggerated. These studies indicate that bird kills from large commercial wind turbines pale in comparison to deaths from several common sources, among them domestic cats, electric transmission lines, windows, pesticides, motor vehicles, and communication towers (Table 1.1). Worldwide, hundreds of millions of birds—perhaps even billions—are killed each year by these sources. Commercial wind turbines, on the other hand, kill a miniscule number of the birds. So why has wind gotten such a bad reputation?

Wind turbines got a bad rap from one of America's oldest and largest wind farms: the Altamont Pass Wind Farm in California. Located just east of San Francisco, Altamont Pass was once home to a mind-boggling 5,000 wind turbines. It is also the habitat of numerous raptors. Soon after the wind turbines were erected, the birds began to perch on the wind towers in search of prey (ground squirrels and other rodents) that live year-round in the grasses at the base of the towers. Some died as they flew into the blades of the turbines toward prey on the ground.

A two-year study of bird kill in the region revealed only 182 dead birds in that time. While any raptor death is of concern to those of us who cherish wildlife, the death rates at Altamont from wind turbines, while significant locally, are insignificant compared to deaths from other factors.

Cats are probably the most lethal force that birds encounter. According to one study, a feral cat kills as many birds in one week as a large commercial wind turbine does in one to two years. Declawing a cat doesn't seem to help much. According to one researcher, the majority of cats (83 percent) kill birds; even declawed and well-fed cats prey on wild birds. Neutering or spaying a cat does not seem to cut down on hunting, either. With more than 64 million cats in America alone, what's the total loss?

No one knows for sure, but if the situation in Wisconsin is indicative of the national toll, America's bird population is being decimated by our furry feline companions. In Wisconsin alone, researchers estimate that cats kill approximately 39 million birds per year. Nationwide, the number is estimated to be around 270 million, and is very likely much higher. "Even if wind were used to generate 100 percent of US electricity needs, at the current rate of bird kills, wind would account for only one of every 250 human-related bird deaths," which includes kills from our furry friends, the cat, according to the AWEA.

Another 130 to 174 million birds die each year as a result of collisions with or electrocution by electrical transmission lines that crisscross the nation. Many

Table 1.1. Estimated Annual Bird Death in the United States by Source

Activity/Source of Bird Mortality	Estimated Annual Mortality
Killed by cats	270 million or more
Collisions with and electrocution by electrical transmission wires	130–170 million
Collisions with windows	100–900 million
Poisoning by pesticides	67 million
Collisions with motor vehicles	60 million
Collisions with communications towers	40–50 million

victims are raptors, waterfowl, and other large birds, electrocuted when their wings bridge two live wires.

Another 100 million to 900 million birds perish after flying into windows, mostly in rural areas, according to another report.

Pesticides kill an estimated 67 million birds each year. And, according to the American Wind Energy Association's report "Facts about Wind Energy and Birds," scientists estimate that about 60 million birds die each year in the United States after being struck by motorized vehicles.

Yet another 40 to 50 million birds perish after flying into communications towers and the guy wires that support them. Studies of one television transmitter tower in Eau Claire, Wisconsin, showed that it killed over 1,000 birds a night on 24 consecutive nights. This same tower killed a record 30,000 birds one evening! A similar tower in Kansas killed 10,000 birds in a single evening.

Another 1.25 million die as a result of collision with tall structures such as buildings, smokestacks, and towers.

Clearly, even the Altamont Pass Wind Farm is benign compared to this host of other factors. Altamont is also an isolated case. No other wind farm in the United States experiences mortality rates remotely close to Altamont Pass. Why?

Aware of the problem, contemporary wind site developers conduct bird studies of potential sites and have been selecting sites for new wind farms that are out of migratory pathways. Improvements in the design of commercial wind turbines have also helped to minimize bird kills at commercial wind farms. Over the years, wind turbines have gotten taller, blades have gotten longer, and the speed at which the blades rotate

has declined substantially. These large, slower-moving blades are more easily avoided by birds, especially raptors. Many of the wind turbines at Altamont Pass have been replaced with these more-raptor friendly models. Ever-larger commercial wind turbines currently under development could reduce the risk even more. Modern wind turbines are also mounted on large tubular towers that, unlike earlier lattice towers, provide no space for predatory birds to perch and watch for prey.

As shown by the graph on the following page, large commercial wind turbines are only a minor source of bird mortality.

"Double the number of turbines," Sagrillo notes, "and we're up to 0.02 to 0.04 percent. Increase the number by 1,000 percent and we're up to 0.1 percent! How many birds does habitat destruction such as mountain top removal to mine for coal kill—forever?!?"

A fact that's rarely discussed is that producing electricity from wind turbines rather than from conventional sources—which create many environmental problems, including air and water pollution, mercury emissions, habitat destruction, and climate change—can actually *benefit* birds and bats—all while protecting a host of other species, ourselves included.

To learn more about efforts to further reduce bird deaths, check out "Facts about Wind Energy and Birds" at the American Wind Energy Association's website, awea.org. Mick has written several articles on the topic, which you can find at renewwisconsin.org. On the lower left, click on "Small Wind Toolboxes." If you want more recent findings, check out the online publication "Wind Turbine Interactions with Birds, Bats, and their Habitats," published in the spring of 2010 by nationalwind.org.

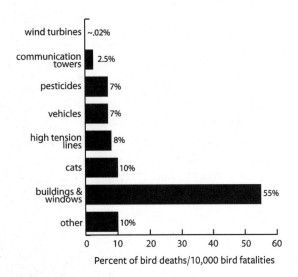

FIGURE 1.10. This graph shows the relative number of bird death from various sources. Note that wind turbines are responsible for only a tiny portion of total annual bird deaths. Credit: Wallace P. Erickson; Western EcoSystems Technology, Inc.

Studies also show that while bats are killed by large commercial wind turbines in certain locations, such occurrences are rare. Most deaths occur during the late summer and early fall, when many species of bats begin their annual migration. Researchers believe that bat deaths occur at this time because the bats may switch off their echolocation to conserve energy while migrating.

According to researchers, large wind turbines in certain locations kill, on average, 2.45 to 3.21 bats per year.

While bat deaths, like bird deaths, are regrettable, there's no indication that bat populations in the vicinity of large or small wind turbines are in any way threatened by them. Other factors play a much larger role in bat mortality, including pesticides, habitat destruction, lighthouses, communication towers, power lines, fences, and human disturbance during hibernation. For more on this topic, check out the accompanying textbox.

Aesthetics

Another downside of wind turbines is that some people don't like the look of them. They believe that wind turbines detract from natural beauty. Although some individuals object to the sight of a residential wind turbine or a commercial wind farm, others find them to be things of great beauty. Ironically, those who find wind turbines to be unsightly often ignore the great many forms of visual blight that litter our landscape, among them cell phone towers, water towers, electric transmission lines, radio towers, and billboards. To be fair, there are differences between a wind tower and these common sources of visual pollution. For one, a turbine's spinning blades call attention to themselves. Another is that they are *new*; we've grown so used to the ubiquitous electric lines and radio towers that many of us fail to see them anymore. Yet another difference is that these structures we choose to ignore are often ones that were erected without public consent. That is, they were exempt from public hearings. People had no choice in their placement and have eventually grown used to them.

Given the opportunity to oppose a structure in their view shed—for example, at a public hearing required for permission to install a residential wind system—many neighbors are quite willing to speak up in opposition. As Mick Sagrillo points out in an article on aesthetics in AWEA's *Windletter*, "anyone who has tried to deal with aesthetics in a public hearing knows only too well why art has never been created by committee."

While the battle continues over commercial wind development and individual battles arise as homeowners or business owners attempt to install small-scale wind to meet their needs, it may be comforting to those who support wind to learn that when windmills were first introduced into Holland, they were looked upon by some with distaste. Another case in point: a large commercial-sized wind turbine built at the Portsmouth Abbey School in Rhode Island drew criticism at first, but is now widely loved by most residents of the community. In Chapter 10, I'll discuss ways individuals can help prevent and overcome opposition from neighbors.

Proximity

Critics raise concerns when it comes to the placement of wind turbines near their properties. Most of the issues related to proximity have been raised by individuals and groups that oppose large commercial wind farms. They cite aesthetics, noise, interference with satellite TV transmissions, potential health issues, and a potential decline in property values.

That said, residential systems can also cause a stir among neighbors. They raise similar issues as well as others, including safety.

An Opposing View

"I think that the goal should be to site wind turbines near homes without fear," Ian argues. "It's better to build relationships with your neighbors, and get them excited about renewable energy. Then the question will be 'Will I be able to see the blades spin?' not 'Must I look at it?' When dealing with neighbors, remember that many of them share your excitement for renewable energy. Many people would love to reduce their electric bill or achieve greater energy independence, or just enjoy the coolness factor that owning a wind turbine brings, but not all can. "Your neighbors can be cool by association," Ian notes, "if you do the PR in advance."

Remember that sound is very subjective—what to some might be irritating is the pleasant sound of renewable energy at work to others.

Ian Woofenden

To avoid problems, we recommend installing turbines, whenever possible, in locations out of sight and hearing of sensitive neighbors. Safety issues like tower collapse are extremely rare and are always the result of bad design or improper installation; should the worst happen though, homeowner's insurance should cover damage to individuals and property. Nonetheless, it is best to place a wind turbine and tower away from your neighbors' property lines to overcome potential objections. In the next few pages, I'll discuss these and other related issues.

Unwanted Sound

Opponents of wind energy and apprehensive neighbors sometimes voice concerns about unwanted sound, aka noise, from residential wind turbines. Small wind turbines do produce sound, and, as wind speed increases, sound output increases.

Sound is produced primarily by the spinning blades and alternators. The faster a turbine spins, the more sound it produces. Over the years, I've found that sound levels vary considerably from one wind turbine to the next. My Skystream 3.7 produces a considerable amount of sound. The 5-kW Endurance wind turbine I installed at a business in Winfield, Missouri, is as quiet as a kitten.

As I point out in Chapter 5, high-rpm wind turbines tend to be louder than low-rpm units. Individuals can reduce unwanted sound by selecting quieter, low-rpm wind turbines. If you are concerned about sound, make this a high priority as you shop for a turbine—and let your neighbors know this is an issue to which you are sensitive.

Wind turbines also come with governing mechanisms, systems that slow down the turbines, even turn them off, when winds get too strong to protect them from damage. Different governing systems result in different sound levels. (I'll discuss this topic in Chapter 5.) When researching your options, I recommend that you listen to the turbines you're considering buying in a variety of wind conditions, including those that require governing. (That's a lot easier said than done, I know.)

Besides buying a quieter wind turbine to reduce sound, it's also important to mount your turbine on a tall tower. Suitable tower heights, which I'll discuss later, are usually at least 80 to 120 feet. (To give you a visual: A 12-story apartment complex would be about 120 feet tall. An 80-foot-tall apartment building would be about 8 stories.) A residential wind turbine mounted high on a tower catches the smoother and stronger—and hence most productive—winds. This strategy also helps reduce sound levels on the ground. Part of the reason for this is that sound dissipates quickly over distance. (For mathematically inclined readers, sound decreases by the square of distance.)

Residential (and commercial) wind turbines are also much quieter than many people suspect because the sounds they make are partially drowned out by ambient sounds on windy days. Rustling leaves and wind blowing around one's ears often drown out some of the sound produced by a residential wind turbine. Experience has shown me that we get used to the sound of wind turbines, too. I wasn't at all pleased when I first installed my Skystream wind turbine in 2010. I could hear it at most locations on my 60-acre property. Now, I barely notice it, and when I do, it brings a smile to my face because I know it's producing electricity from the strong winds that blow through.

Sound is measured in two ways—by loudness and by frequency. Loudness is measured in decibels (dB). Frequency is the pitch. A low note sounded on a guitar has a low frequency or pitch. A high note has a high frequency. Let's consider loudness first.

We live in a noisy world. In fact, the average background noise in a house is about 50 dB. Nearby trees on a breezy day measure about 55 to 60 dB. According to Sagrillo, "Most of today's residential wind turbines perform very near ambient levels over most of their effective operating range."

However, even though the intensity of a sound produced by a wind generator may be the same as ambient sound, the frequency often differs. As a result, wind turbine sounds may be distinguishable from ambient noises, even though they are not louder. The blades of my Skystream, for instance, produce a high-pitched sound when the winds blow at high velocity. It's quite noticeable.

"Today's home-sized wind turbines typically operate from just below to just above ambient environmental sound levels at their loudest when governing," Sagrillo notes. "This means that while the sound of a wind turbine can be picked out of surrounding noise if a conscious effort is made to hear it, home-sized wind turbines are by no means the noisy contraptions that some people make them out to be."

For more on sound, you may want to read Mick's column, "Residential Wind Turbines and Noise" in the April 2004 issue of AWEA's *Windletter*. I'll also spend more time on this topic in Chapter 5, and in Chapter 10 I'll discuss strategies for addressing sound issues at zoning hearings.

Ice Throw

Like trees and powerlines, wind turbines can ice up during ice storms. Some worry that ice buildup will result in "ice throw"—hunks of ice being hurled off the blades. This, they worry, may pose a danger to people and structures on the ground.

While ice builds up on blades in ice storms, it is typically deposited on turbines and towers in very thin sheets. When the blades are warmed by sunlight, however, the ice tends to break up into small pieces, not huge and potentially dangerous chunks. Moreover, when it melts, it falls down, around the base of the tower. It's not flung into neighbors' yards.

Also bear in mind that ice buildup on the blades of a wind turbine dramatically reduces the speed at which a turbine can spin.[1] Ice accumulation on a wind turbine blade is a little like trying to drive a car with four flat tires; there is a lot of resistance to movement. Because the turbine is slowed down so much, ice is not thrown great distances; it tends to fall around the base of the tower—just as it does from trees and power lines.

Any prudent person would be advised to stay away from the tower base when ice is shed from the blades, just as they would from ice falling from trees or power lines. Ice-laden trees are also considerably more dangerous, as ice-coated branches can and often do break and fall to the ground, damaging power lines and cars or houses. Entire trees can topple as a result of ice buildup.

On the rare occasion that ice builds up on a wind turbine, experienced wind turbine operators shut down their turbines until the Sun or warmer temperatures melt the ice; no electricity is generated when ice induces such low rpms anyway. By the way, we've experienced numerous ice storms over the past six years and never once had to worry about ice dropping from or being flung by our wind turbine.

Interference with Telecommunications

Some opponents of wind energy also raise the issue of interference with telecommunications signals.

While there are a few reports of large-scale wind turbines causing interference with television reception, these problems arose because the turbines were installed directly in the line of sight between the TV transmitter and a residential antenna. The spinning blades chopped up the signal, causing flickering on televisions. Interference represented isolated cases and was easily corrected by installing larger antennas or signal boosters.

With small wind turbines, interference is extremely unlikely. Turbines for homes and small businesses have small blades that do not interfere with such signals. The blades of modern wind turbines are also made out of materials that are unlikely to cause problems. Unlike the metal blades of years past, which can reflect TV signals, the fiberglass and plastic blades in use today are "transparent" to telecommunications signals.

As a case in point, we should note that small wind turbines are often installed to power remote telecommunications sites. A US-based small-wind turbine manufacturer, Abundant Renewable Energy (now operating as Xzeres Wind Power), mounted their internet receiver/transmitter on their wind turbine tower. Telecommunication equipment wouldn't be installed in such locations if there were problems with interference.

Strobing and Photosensitive Epilepsy

Yet another issue that may be raised from time to time by concerned neighbors or opponents of wind energy is the possibility that shadow flicker from wind turbines will stimulate epileptic seizures in individuals who suffer from photosensitive epilepsy. This is an extremely rare type of epilepsy in which seizures are triggered by flickering or flashing light.

This concern, most often raised by opponents of large commercial wind turbines, but occasionally raised at zoning hearings for small wind turbines, is not just overblown, it's not even true. According to researchers, there's never been a case of epileptic seizure triggered by a wind turbine in human history.

While blades of small turbines may form small and rather vague shadows, it is difficult to see the shadow of individual blades due to the speed with which the blades spin. As Jim Green notes, "The rotors of residential-scale wind turbines, 10 kilowatts and smaller, essentially become transparent at typical operating speeds because the blades spin faster than the eye can detect."

The true test of this issue's seriousness may come from installers and dealers. Mike Bergey of Bergey WindPower, for example, has never received a complaint about shadow flicker from customers or neighbors, and he's been in the business for nearly 39 years, as of 2016.

Property Values

Opponents of wind farms often raise the specter of declining property values, despite the lack of any evidence to support their assertions. Nonetheless, concerns over property values often arise in zoning hearings over small wind turbines. As Sagrillo puts it, the rationale is that the neighborhood "view shed" will be compromised as a result of the installation of a home-sized wind turbine. Neighbors worry that they will not be able to sell their property for its true value.

While wind turbines on tall towers are visible, lots of other tall structures like silos, barns, high-power transmission lines, water towers, and cell phone towers are present in the same rural environments where residential and small business wind systems are typically installed. Small wind systems are often much less visible than these structures. Moreover, I've never heard of an instance in which a residential wind turbine adversely affected the value of a neighbor's property. For the system owner, a wind turbine could increase property values, in part as a result of reduced utility bills. I find my wind turbine perched on top of the hill to be a thing of beauty. You can easily see it from a mile away—and it looks terrific! In fact, I often use it to help people locate my educational center. "Turn left onto Pin Oak Road. We're a mile up on your left. You'll see a wind turbine on a tower on top of the hill. That's our place," is what I tell students and delivery truck drivers.

The Advantages of Wind Energy

Although residential wind turbines and their energy source, the wind, have their downsides, many features make them well worth considering. To begin with, wind energy is an abundant and renewable resource. We won't run out of wind for the foreseeable future—which stands in stark contrast to the future of coal, natural gas, oil, uranium.

Wind energy—both large and small—can also play a meaningful role in offsetting declining US natural gas supplies. In the United States, 33 percent of all electricity is currently generated by natural gas, according to the U.S. Department of Energy. As supplies continue to decline, wind could help ease the crunch, supplying a growing percentage of our nation's electrical demand long into the future.

Wind could even replace nuclear power plants the world over. Nuclear power plants generate about 20 percent of the United States' electricity, and substantially higher percentages in countries such as France. Although wind energy does have its impacts, it is a relatively benign technology compared to fossil fuel and nuclear power plants. Because of this, it could help all countries create a cleaner and safer energy future at a fraction of the cost and impact of conventional electrical energy production.

While many proponents of wind and solar energy claim that these technologies will help us reduce our dependence on oil, few realize that in the United States only a tiny fraction (about two percent) of our electricity comes from burning fuels derived from

FIGURE 1.11. Dan and His All-Electric Nissan Leaf. This is a perfect commuter car. Depending on the model, it travels 84 to 104 miles on a single charge. If you are replacing a car that gets 30 miles per gallon, and you pay 10 cents per kilowatt-hour for electricity from a utility, it would cost you about 68 cents to travel 30 miles on electricity. The same trip in a gas-powered vehicle would cost $2.00 (if gas prices were $2.00 per gallon). With higher gas prices, the savings accrued when driving an electric car become even more attractive. Credit: Linda Stuart.

oil. That said, if wind and solar-generated electricity were used to power electric vehicles, slowly but surely replacing gasoline-powered vehicles, these renewable energy sources could help us reduce our dependence on oil. I drive a Nissan Leaf powered entirely by solar and wind energy. In the first six months, it reduced our consumption of gasoline by 90 gallons—and our other car is a Toyota Prius that routinely gets 50 miles per gallon or more.

Another huge benefit is that when wind is used to power electric and plug-in hybrid vehicles, it helps them clean up the air and reduces the buildup of the greenhouse gas carbon dioxide, which is largely responsible for global warming and climate disruption.

Another benefit of wind energy is that, unlike oil, coal, and nuclear energy, the wind is not owned by major energy companies. The cost of wind is not subject to price increases. Price hikes caused by rising fuel costs are not probable in a wind-powered future. However, this is not to say that wind energy is immune to the rising price of fossil fuels. While the fuel itself (the wind) will not increase in price, the price of wind generators is likely to increase as traditional fuel prices rise. That's because it takes energy to extract and process minerals to make the steel, copper, and rare earth magnets needed to manufacture wind turbines and towers. It also takes energy to make turbines and towers and ship and install them.

An increasing reliance on wind energy—and electric cars—could also ease political tensions worldwide. If we free ourselves from Middle Eastern oil, we won't need costly military operations aimed, in part, at stabilizing a region where the largest oil reserves reside. We'll likely never fight a war over wind energy resources. Not a drop of human blood need be shed to ensure a steady supply of wind energy to the fuel the economy.

Yet another advantage of wind-generated electricity is that it uses existing infrastructure, the electrical grid, and existing technologies like computers, microwaves, and so on. A transition to wind energy could occur fairly seamlessly.

Individuals can also meet all or part of their energy needs in rural areas with good wind resources at rates that are competitive with conventional electricity. In remote locations, wind or wind and solar-electric hybrid systems may be cheaper than conventional power delivered through newly installed and costly electric lines from the utility grid.

Finally, lest we forget, wind is a clean resource. Wind energy will help homeowners and businesses do their part in solving costly environmental issues such as acid rain and global climate change. As Mick points out in his workshops, the average home in the United States consumes 900 kilowatt-hours of electricity per

Envision a future in which distributed (small) wind power is *embraced* in the local landscape because it *expresses community support* for clean air, reduced carbon emissions, and strong local economies through use of a sustainable, indigenous energy source.

Jim Green, National Renewable Energy Laboratory

month. Replacing the electricity generated by a coal-fired power plant with wind-generated electricity will reduce a family's consumption of coal by approximately 5.5 tons per year. This, in turn, will reduce the emission of carbon dioxide by about 11 tons per year. It will also reduce mercury emissions. You couldn't ask for more reasons to justify a switch to wind energy.

Wind energy also provides some substantial economic benefits. Wind energy development creates jobs—more jobs per kilowatt-hour generated than any other type of power plant. At the start of 2016, American wind power supported a record 88,000 jobs—an increase of 20 percent in a year—according to AWEA's *U.S. Wind Industry Annual Market Report* for the year ending in 2015. "Strong job growth, the report notes, "coincided with wind ranking number one as America's leading source of new generating capacity last year, outpacing solar and natural gas."

Wind energy development also concentrates economic benefits locally and within states. "Wind power benefits more American families than ever before," notes Tom Kiernan, CEO of AWEA. "We're helping young people in rural America find a job close to home. Others are getting a fresh chance to rebuild their careers by landing a job in the booming clean energy sector. With long-term, stable policy in place, and a broader range of customers now buying low-cost wind-generated electricity, our workforce can grow to 380,000 well-paying jobs by 2030."

And wind power does not require extensive use of water, an increasing problem for coal, nuclear, and gas-fired power plants, particularly in the western United States and other drought-stricken areas.

The Purpose of This Book

This book's principal focus is on small wind-electric systems. As noted earlier, the rated output of small wind turbines ranges from 1 kilowatt to 100 kilowatts. Most of the turbines I'll be discussing fall in a narrow range from 1 kilowatt to 20 kilowatts. The blades of small turbines (1 to 100 kilowatts) run from 4 feet to 32 feet in length. Small-scale wind systems serve a variety of purposes. The smaller units within this category are sufficient to power cabins and cottages. Larger turbines power homes, small businesses, schools, farms, ranches, manufacturing plants, and public facilities.

This book is written for individuals who want to learn about small-scale wind systems. It is also written for those who aren't particularly well versed in electricity and electronics. You won't need a degree in electrical engineering, renewable energy, or physics to make sense of the material covered in this book.

The overarching goal in writing this book was to create a user-friendly book that teaches readers the basics of wind energy and wind energy systems. This book is *not* an installation manual. It will not turn you into a wind energy installer or equip you to install a wind turbine and tower on your own. It will, however, help you determine if wind energy is right for you. When you finish reading and studying the material in this book, you'll know an amazing amount about wind and wind energy systems. You will have the knowledge required to assess your electrical consumption as well as the wind resource at your site, and to determine if wind will meet your needs.

When you are done with this book, you should have a good working knowledge of the key components of wind energy systems, especially wind turbines, towers, batteries, and inverters. In keeping with my long-standing goal of creating knowledgeable buyers, this book will help you know what to look for when shopping for a wind energy system. You'll also know how wind turbines are installed and have an understanding of their maintenance requirements.

If you choose to hire a professional wind energy expert to install a system— a route I highly recommend—you'll be thankful you've read and studied the material in this book. The more you know, the more input you will have into your system design, components, siting, and installation—and the more likely that you'll be happy with your purchase.

In keeping with another long-standing goal of mine, this book should also help readers develop realistic expectations. I believe that those interested in installing renewable energy systems need to proceed with their eyes wide open. Knowing the shortcomings of wind energy—or any renewable energy technology, for that matter—helps avoid mistakes and prevent disappointment fueled by unrealistic expectations.

Over the years, I've found that there's a lot of ignorance about wind and its potential, and this leads to very unreal expectations. Many people I've talked to think they can simply install a wind turbine on a short tower on their property and live happily ever after, merrily consuming electricity as if there were no tomorrow. Truth is, you will need very likely need a fairly large wind turbine mounted on a tall tower and a very good (translated: windy) site to make such a dream come true. In addition, wind energy systems require annual inspection and maintenance—climbing or lowering a tower to access the wind turbine to check for loose fasteners and blade damage and, much less commonly, to make an occasional part replacement. If you are not up for it or don't want to pay someone to climb or lower your tower once or twice a year to check things out, you may want to invest in a solar-electric system instead.

Organization of This Book

Now that you know a little bit about the history of wind energy, the pros and cons of this clean, renewable energy source, and the purpose of this book, let's start our exploration. I'll begin in the next chapter by studying wind, the driving force in a wind energy system. You will learn how winds are generated and explore the factors that influence wind flows in your area.

In Chapter 2, I will also explore the factors that affect energy production by a residential wind turbine. That is, you will learn the simple, undeniable mathematics of wind energy. The math isn't difficult, and this discussion will demonstrate how the proper design and placement of a wind turbine can result in dramatic increases in electrical output. When you finish, you will understand why it is important to mount a wind turbine as high as you can and out of the way of obstructions that reduce wind speed and create turbulence when the winds blow. This advice could make the difference between a successful wind venture and a costly failure.

In Chapter 3, I'll explore wind energy systems. You'll learn the three types of residential wind energy systems: (1) off-grid, (2) batteryless grid-tie, and (3) grid-connected with battery backup. You'll also learn about the basic components of each one. We'll also look at hybrid systems.

Chapter 4 explores the feasibility of tapping into wind at your site. I'll teach you how to assess your electrical energy needs and how to determine if your site has enough wind to meet them. You'll learn why cost-effective energy-efficiency measures that reduce your electrical demands will save you heaps of money when buying a wind energy system. You'll also learn ways to evaluate the economics of a wind system.

Chapter 5 introduces you to wind turbines. You'll learn about the different types of wind turbines and how they work. I'll also give you shopping tips—what to look for when buying a wind turbine. I'll even spend a little time looking at ways to build your own wind generator and introduce you to wind turbines designed to pump water.

Chapter 6 describes the three basic tower options. You will learn how towers and guy wires (used to support certain types of towers) are anchored to the ground. I'll underscore the importance of mounting a wind turbine high above the ground—out of turbulent ground-level air and dead air zones and into the much smoother and more powerful winds that blow higher up. I'll also look briefly at how towers are installed.

Chapter 7 addresses another key component of wind energy systems, the inverter. You will learn how inverters work, what functions they perform, and what to look for when shopping for one.

In Chapter 8, I'll tackle storage batteries, one of the key components of off-grid wind systems. You will learn whether you will need a battery bank and, if so, what kind of batteries you should install. You will learn about battery care and maintenance and ways to make your life with batteries much easier. You'll benefit from my many years of experience with battery systems as I point out common mistakes and ways to avoid them. You will also learn about battery safety and how to size a battery bank for a wind energy system.

In Chapter 9, I'll give a brief overview of wind energy system maintenance. Because each wind turbine and tower is different, I won't go into specifics. I will, however, stress the importance of regular inspection and maintenance and describe some of the things you'll have to do to keep your wind energy system performing optimally.

With this information in mind, in Chapter 10 I'll explore a range of issues such as homeowner's insurance, financing renewable energy systems, obtaining building permits and electrical permits, and zoning issues.

Finally, this books ends with a fairly comprehensive Resource Guide. It contains a list of books, articles, videos, associations, organizations, workshops, and websites on residential wind energy.

What do you say—shall we get started?

CHAPTER 2

Understanding Wind and Wind Energy

Wind can be an asset and a liability. Growing up along the windswept shores of Lake Ontario in western New York, fierce winter winds rattled the aluminum storm windows of my bedroom, waking me in the middle of the night. Other times, the winds created nightmarish blizzards that made driving impossible. One night, driving my girlfriend home in a blizzard I drove the family car into a ditch.

Wind has proved to be a liability in my adult years, too. On several kayak trips on Western rivers like the Colorado and Green Rivers, the winds were so strong that they flipped kayakers over in relatively calm water. When kayaking the Grand Canyon, powerful winds sometimes halted the forward motion of the support rafts that carried our gear. Several of us had to climb aboard the rafts to help the oarsmen battle the winds. And on several of my backpacking trips in the high peaks of Colorado, powerful winds have made travel across the tundra nearly impossible. Sometimes, it seemed as if we'd be blown off the mountaintops.

Wind has also been a blessing. Those powerful winter winds off Lake Ontario, for instance, created huge snowdrifts along the hedgerow that bordered my family's property. My brothers and I would dig into the deep drifts with snow shovels, creating a network of tunnels where we could hang out in safety, protected from the fierce winds.

Today, wind continues to bless my life. It provides a portion of the electricity to my home, farm, and educational center in east central Missouri. This chapter will introduce you to the wind—the driving force of a wind energy system. I'll examine how wind is generated and study four distinct types of wind: (1) offshore and onshore winds, (2) mountain-valley winds, (3) global circulation patterns, and (4) storms. I'll also explore ways local topography affects wind, introducing you to two key concepts in wind energy: ground drag (caused by friction) and turbulence. Chapter 4 will cover more details on wind, including the influence of topography and vegetation on wind.

As you read this material, bear in mind that this is not just a theoretical exercise. It provides the practical knowledge you will need to select the best site for a wind turbine and the optimum tower height.

What Is Wind?

Wind is air in horizontal motion across the Earth's surface. All winds are produced by differences in air pressure between two adjoining regions. Differences in pressure result from differential heating of the surface of the Earth, as you shall soon see. Heating, of course, is the work of the Sun.

Like most other forms of energy in use today, even coal, oil, and natural gas, then, wind is a product of sunlight. Some wind advocates, in fact, refer to wind as "the *other* solar energy" (the same is said about hydropower because the hydrologic cycle is also driven by the Sun). Some individuals refer to wind as "secondhand solar energy." Paul Gipe, an author of several books on wind energy, moderated a workshop at Solar 2007, the annual conference of the American Solar Energy Society appropriately titled "Bringing 'Solar Convection' Technology into the Mainstream: Small Wind Turbines." Convection, of course, is the movement of air from one area to another because of differences in air temperature and hence pressure, a topic I'll discuss shortly.

To understand wind, we begin by looking at two types of local winds: (1) offshore and onshore winds, and (2) mountain-valley breezes.

Offshore and Onshore Winds

Offshore and onshore winds are, as their names imply, generated along coastlines. That is, they occur along the shores of large lakes, such as the Great Lakes of North America, and along the coastlines of the world's oceans. Offshore and onshore winds blow regularly, nearly every day of the year. Like all other types of wind, solar energy generates these winds. How?

As shown in Figure 2.1a, sunlight shining on the Earth's surface heats the land and nearby water bodies. As the water and adjoining land begin to warm, they radiate some of the heat (infrared radiation) into the atmosphere. This heat, in turn, warms the air immediately above them. When air is heated it expands, and as it expands it becomes less dense. As the density of the air decreases, it rises. The upward movement of air is called a thermal or updraft.

Although water and land both heat up when warmed by the Sun, land masses warm more rapidly than neighboring bodies of water. Because air over land heats up more quickly than air over water, air pressure over a landmass is lower than it is over neighboring surface waters.

Low-pressure air rising from the land draws in cooler, higher-pressure air lying over neighboring water. As the cool air flows in to fill the void, a breeze is created. The breeze is known as an *onshore breeze* or *onshore wind*.

At night, the winds blow in the opposite direction—that is, they blow from land to water, as illustrated in Figure 2.1b. These are known as offshore breezes or offshore winds.

Like onshore winds that occur during the day, offshore winds are created by differences in air pressure between the air over land and neighboring water bodies. Differential cooling generates these differences. Here's what happens: after sunset, the land and the ocean both begin to cool. However, land gives up heat more rapidly than water. As a result, the land surfaces cool more rapidly than nearby water bodies. As shown in Figure 2.1b, at night the air over water continues to release heat, causing the air above it to expand and rise. As the warm, low-pressure air rises, it creates a slight updraft over the water. Cooler, higher-pressure air from land flows in to fill the vacuum. The result is an offshore breeze: steady winds that flow from land to water.

In the absence of prevailing winds or storms, discussed shortly, offshore and onshore breezes operate day in and day out on sunny days, providing a steady supply of wind energy to power wind turbines. Because offshore and onshore winds are fairly reliable, coastal regions of the world are often ideal locations for wind turbines.

Coastal winds are a great energy resource and could help many countries meet their needs for electricity. In fact, it is not unusual to encounter winds with an average annual speed of 12 to 18 miles per hour (5 to 8 meters per second) in such locations. Wind speeds such as these are ideal for residential and commercial wind

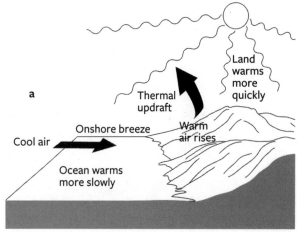

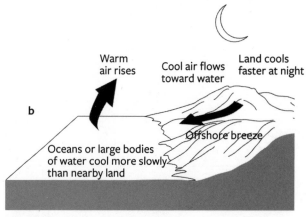

FIGURE 2.1. Onshore and Offshore Breezes. Onshore (*a*) and offshore (*b*) breezes occur along the coastlines of major lakes and oceans. Credit: Anil Rao.

generators, although the onshore and offshore wind effect is restricted to within a mile or so from the coastline.

Coastal winds are not only more consistent than winds over the interior of continents, they also tend to be more powerful because of the relatively smooth and unobstructed nature of the surface of open waters. Put another way, wind moves rapidly over water because lakes and coastal waters provide very little resistance to its flow, unlike forests or cities and suburbs, which dramatically lower surface wind speeds for reasons explained shortly.

Mountain-Valley Breezes

Like coastal winds, mountain-valley breezes arise from the differential heating of the Earth's surface. To understand how these powerful winds are formed, let's begin in the morning.

As the Sun rises in the eastern sky on clear days, sunrays strike the valley floor and begin heating the ground, valley walls, and mountains. As the ground and valley walls begin to warm, the air above them warms. It then expands and begins to flow upward. This process is known as convection. (Convection is the transfer of heat in a fluid or air that is caused by the movement of the heated air or fluid itself. In a building space, warm air rises and cold air settles to create a convection loop.) While some of this air rises vertically, mountain valleys also tend to channel the solar-heated air up toward the mountains (Figure 2.2). As the warmed air moves up a valley, cooler air from surrounding areas such as nearby plains flows in to replace it. The wind flowing into and up the valley is known as a *valley breeze.*

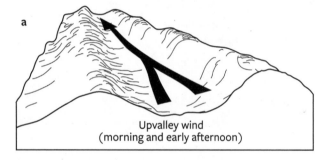

a

Upvalley wind
(morning and early afternoon)

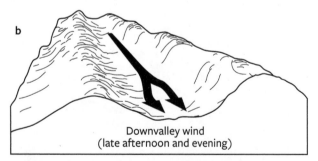

b

Downvalley wind
(late afternoon and evening)

FIGURE 2.2. Mountain-Valley Breezes. Mountain-valley winds can provide a reliable source of wind power if conditions are just right. (*a*) Morning up-valley winds. (*b*) Evening down-valley winds. Credit: Anil Rao.

Throughout the morning and well into the afternoon, breezes flow up-valley—from the valley floor into the mountains. These breezes tend to reach a crescendo in the afternoon. When the Sun sets, however, the winds reverse direction, flowing down valley.

Winds flow in reverse at night because the mountains cool more quickly than the valley floor. Cool, dense air (high-pressure air) from the mountains sinks and flows down through valleys like the water in a mountain stream, creating steady and often predictable down-valley or mountain breezes.

Mountain breezes typically start up shortly after sunset and continue throughout the night, terminating at sunrise. As a rule, mountain winds are generally strongest in valleys that slope steeply and valleys cradled by high ridges. The strongest mountain winds occur at the mouth of the valley (that is, the lower reaches). The winds in the upper reaches of the valley, however, tend to be smoother and less turbulent than the winds in the lower parts.

Together, valley and mountain winds are known as mountain-valley breezes. As a rule, mountain breezes (down-flowing winds) tend to be stronger than daytime valley breezes. Mountain breezes often reach speeds around 25 miles per hour (about 11 meters per second), ideal for small and large-scale wind energy production. Wind flow is most rapid, as a rule, in the center of the valley—at about two-thirds the height of the surrounding ridges.

Mountain-valley breezes typically occur in the summer months, when solar radiation is greatest. They also typically occur on calm days when the prevailing winds (larger regional winds, discussed shortly) are weak or nonexistent.

Mountain-valley winds may also form in the presence of prevailing winds—for example, when a storm moves through an area. In such instances, mountain or valley winds may "piggy back" on the prevailing winds, creating even more powerful (and hence higher energy) winds. When consistently flowing in the same direction, such winds can provide a great deal of power that can be tapped to produce an abundance of electricity.

When assessing a wind site in a mountain valley, the orientation of the valley to the prevailing wind direction is extremely important. If the prevailing winds run in the same direction as the valley floor, they can enhance the wind resource—and could dramatically increase the amount of electricity that can be generated by a wind turbine.

When looking for land to build on and install a wind turbine, shop carefully in mountainous terrain. The ideal property in such terrain is a valley that runs parallel to the prevailing wind in the summer, when mountain-valley winds are strongest. As just noted, in such instances, mountain-valley winds can piggyback on the prevailing winds, creating more harvestable wind power. In the winter, when the mountain-valley breezes are the weakest, prevailing winds will continue to flow through the valley, spinning the blades of a wind turbine. (Bear in mind that winter and summer winds may come from different directions.)

As a rule, "a valley that is both parallel to the prevailing wind and experiences mountain-valley winds will provide sites that are dependable sources of power," according to the authors of *A Siting Handbook for Small Wind Energy Conversion Systems*. If the valley narrows at a site, the narrowing creates a funnel effect (aka the Venturi effect) that further increases wind speed—and the potential output of a wind generator.

If the prevailing winds run perpendicular to the valley floor, or at an angle greater than 35 degrees, expect much less electrical energy from a wind turbine. Such winds tend to flow over valleys. The deeper and narrower the valley, the more likely cross-valley winds will flow over it.

Cross-valley winds may also combine with solar-heated air rising from the valley floor to create eddies—swirling turbulent air. As you shall soon see, turbulence wreaks havoc on wind turbines.

When considering a wind site in a valley, bear in mind that not all valleys are created equal. Valleys that do not slope downward from mountains are not usually good wind sites; they can be, though, if they're in regions with extremely strong winds or they narrow significantly, funneling prevailing winds through them. (The Columbia Gorge is a good example.) If you're thinking about installing a wind turbine in a valley not connected to a chain of mountains, you'd probably do better placing it on a surrounding ridge.

Large-scale Wind Currents

Local winds like those just discussed can be a valuable source of energy for homes and businesses powered by wind generators. However, the winds on which most people rely are those derived from much larger air flows resulting from regional and global air circulation. These create dominant wind flow patterns, known as prevailing winds.

Prevailing winds, like local winds, are created by the differential heating of the Earth's surface, but on a much larger scale. Here's a simplified version of how they are created. As shown in Figure 2.3, the Earth is divided into three climatic zones: the tropics, temperate zones, and poles. As illustrated, the tropics straddle the equator, extending from 30° north latitude to 30° south latitude. Because the tropics are more directly aligned with the Sun throughout the year than the other zones, they receive considerably more sunlight. As a result, tropical regions are the warmest places on Earth.

The temperate zones lie outside the tropics, both in the Northern and Southern Hemispheres. They are located between 30° north latitude and 60° north latitude in the Northern Hemisphere and 30° south and 60° south latitude in the Southern Hemisphere. They receive less sunlight than the tropics and are, therefore, noticeably cooler. Last but not least are the North and South Poles. They receive the least amount of sunlight and are the coolest regions of our planet, and historically have been covered by ice much of the year. (That's changing because of global warming.)

The unequal heating of the Earth's surface is responsible for the movement of huge air masses across its surface. As shown in Figure 2.4a, a highly simplified view of solar convection, hot air produced in the tropics rises upward. Cool air from the northern regions—as far north as the poles—moves in to fill the void. The result is huge air currents that flow from the poles to the equator.

Although air generally flows from the North and South Poles toward the equator, circulation patterns are much more complicated. In the Northern Hemisphere, for instance, warm air from the equator rises and moves northward. As shown in Figure 2.4b, as the warm tropical air, which is laden with moisture, flows northward it begins to cool, depositing moisture. (Most of the moisture falls back on the tropics, being locally recycled.)

As shown in Figure 2.4b, some of the cooler northbound air sinks back to the Earth's surface and flows back toward the equator, creating a short-circuit in global wind circulation. This results in winds that blow toward the equator, known as the trade winds.

Because the trade winds blow quite consistently, day after day, they are a potentially huge and reliable source of energy, a fact not missed by sailors who have relied on them for centuries. (The trade winds got their name from their importance to trade in the early days of commerce.) For residents and nations fortunate enough to lie in the trade winds, these constant winds could become a valuable source of electrical energy.

As shown in Figure 2.4b, not all of the air flowing northward from the equator returns via the trade winds. A substantial amount of the

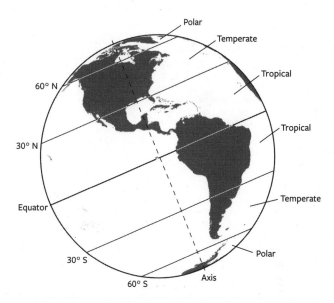

FIGURE 2.3. Climate Zones. The Earth is divided into three climate zones in each hemisphere. In the Northern Hemisphere, warm air from the tropics flows northward by convection, creating the global circulation pattern shown here. Credit: Anil Rao.

Global Air Circulation

The flow of air between the equator and North and South Poles distributes excess heat from equatorial regions throughout the planet. (As a result the equator is cooler and the temperate and polar regions are warmer than they would be without global air circulation.) This massive movement of air is also the source of winds that could supply human society with a significant portion of the electrical energy it needs.

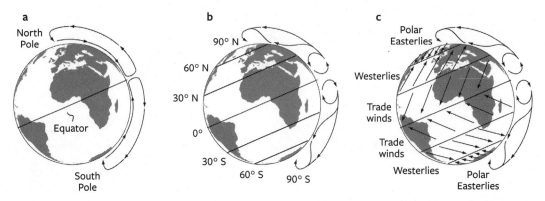

FIGURE 2.4. Global Air Circulation. (*a*) As shown here, warm tropical air rises and flows toward the poles. Cold polar air flows toward the equator. (*b*) Air circulation is more complicated than shown in (*a*), however. Warm tropical air loses some of its heat and sinks, creating a circulation pattern that brings air back toward the equator, creating the trade winds. Air masses moving over the temperate zone split into upper and lower winds. (*c*) Wind patterns caused by the Coriolis effect, resulting from the rotation of the Earth on its axis. Credit: Anil Rao.

northward moving air proceeds toward the North Pole. Devoid of much of its moisture, the relatively dry air travels over the temperate zone.

As Figure 2.4b also shows, the air masses flowing northward across the temperate zone split into two: higher- and lower-level winds. As it flows northward, the lower-level air picks up a considerable amount of moisture, much of which falls as rain and snow over the temperate zone.

When the equatorial air reaches the North Pole, however, it is dry and cold. This cold, dry air mass then sinks and begins to flow southward back toward the equator.

If no other forces were at work, winds flowing back to the equator would flow from north to south. As shown in Figure 2.4c, they don't. Other factors influence the movement of air masses across the surface of the planet. One of the most significant is the Earth's rotation, which results in a phenomenon known as the Coriolis effect, described next.

The Coriolis Effect

In order to understand why prevailing winds deviate from the expected flow patterns based solely on convection, let's look at the Northern Hemisphere once again. We'll focus our attention first on the trade winds. As shown in Figure 2.4c, the trade winds in the Northern Hemisphere flow not from north to south, as you might expect, but from the northeast to southwest. Why?

The quick answer is: because they are "deflected" by the Earth's rotation.

In reality, the Earth's rotation doesn't deflect winds. It just makes it appear as if the winds have been deflected. The apparent deflection in wind direction in the tropics is a planetary sleight of hand, an illusion produced by the rotation of the Earth on its axis. To understand this phenomenon, let's look at a simple example. Imagine that you board a plane leaving the North Pole. The pilot plots a course that will take you due south toward an airfield on Sri Lanka just south of India. If the pilot flies the plane due south the entire trip, will you arrive at your destination in Sri Lanka?

No.

The plane will end up somewhere over the Arabian Sea.

Why?

Because of the Earth's rotation.

As shown in Figure 2.5, as the plane travels south, the Earth rotates beneath it. The Earth rotates eastward. If you plot the flight path of the plane it appears to have been deflected.

It hasn't.

It only looks that way.

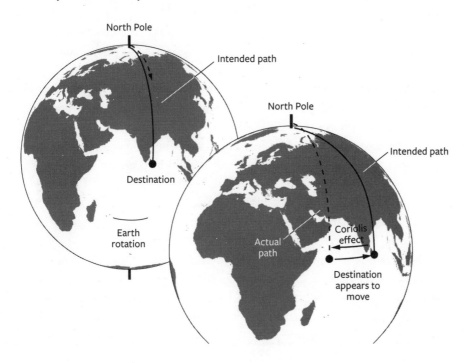

FIGURE 2.5. The Coriolis Effect. The rotation of the Earth causes an apparent deflection in the path of winds. This can be understood by observing the flight of a plane that begins at the North Pole and heads directly south toward Sri Lanka. The plane appears to veer off course. It hasn't. But the Earth's rotation makes it look that way. Credit: Anil Rao.

The apparent deflection of the plane's path is the Coriolis effect, named after the scientist who first described it. In the Northern Hemisphere, the deflection is to the right of the direction of travel. In the Southern Hemisphere, the deflection is to the left.

Winds flowing north or south also appear to be deflected thanks to the Coriolis effect. The trade winds, for instance, appear to flow from northeast to southwest in the Northern Hemisphere.

In the temperate zone, as shown in Figure 2.4c, the low-level north-flowing winds that sweep across the surface of the Earth flow across the Northern Hemisphere not from south to north but from the southwest to northeast. These are the prevailing southwesterly winds that blow across the Great Plains of North America. Many a wind farm and many small-wind-system owners depend on them for electricity. The direction they flow is the result of the Coriolis effect.

Global airflows provide a source of reliable energy that could be available to use for five billion years as long as their source, the Sun, remains. But there's more to wind than that.

Wind from Storms

As you have just seen, wind flow patterns are complicated. They consist of local winds like onshore and offshore winds as well as enormous global air movements. Of course, winds are often associated with storms. Storms, as you probably suspect by now, are produced when high-pressure and low-pressure air masses collide. Let's take a brief (and highly simplified) look at the complex interactions that produce fierce, high-energy winds that can be a blessing and a curse to wind energy systems.

As anyone who has watched the evening news realizes, high and low pressure zones move across the continents, interacting with one another and creating a wide assortment of weather. Where do these high- and low-pressure air masses come from?

Low-pressure zones originate in the tropics.

Encircling the Earth at the equator is a permanent, elongated, narrow band of low pressure. It's referred to as the *equatorial low-pressure trough*.

The equatorial low-pressure trough is produced by the huge influx of solar energy in equatorial regions. This heats water and land masses, which, in turn, warms the air above them, causing it to expand, become lighter, and rise. The atmospheric pressure in the equatorial region is therefore lower than in the Northern or Southern Hemispheres. The equatorial low-pressure region is not stable, however. Huge masses of low-pressure air frequently "break off" from the trough and migrate northward, sweeping across the North American continent.

In contrast, the North and South Poles are regions of more or less permanent cold high-pressure air. (Remember cold air is denser than warm air masses and has a higher atmospheric pressure.) Like warm tropical air, huge masses of cold Arctic air also break loose and drift southward, sweeping across the Northern Hemisphere. In the summer, cold Arctic air masses may bring welcome relief to people beleaguered by blisteringly hot temperatures. In the winter, these high-pressure air masses deliver cold Arctic air as far south as Florida and southern California, freezing pipes and sending people and animals scurrying for shelter.

High-pressure and low-pressure air masses, often measuring 500 to 1,000 miles in diameter, move across continents in a magnificent global ballet. As high- and low-pressure air masses collide with one another, they produce an assortment of exciting, sometime dangerous weather. For example, when cold, dry air masses moving in from the north collide with warm humid air masses from the south, snow and rain form. (Cold air causes the moisture in the warm air masses to condense, resulting in precipitation.)

The movement of high- and low-pressure air masses across continents is steered by prevailing winds and by the jet stream (high altitude winds). Storms produced when these air masses meet are typically characterized by intense winds. Winds are created when air flows from high- to low-pressure air masses. The greater the difference in pressure between a high-pressure air mass and a "neighboring" low-pressure air mass, the stronger the winds. In some cases, these winds contain an enormous amount of energy that can be a blessing to those who rely on wind turbines. They can also, as noted above, be a curse. Really fierce winds can damage wind systems.

It is often assumed that wind turbines will capture all the energy of a storm. Unfortunately, that's not true. Storms are too infrequent and of too-short duration to make it worthwhile to engineer and build a wind generator strong enough to keep spinning and producing electricity in very strong winds. Moreover, a wind turbine engineered to capture the energy of strong storms would not function very well in low-wind speeds, which are much more common.

As you will learn in Chapter 5, all wind generators worth owning have a governing system that slows them down in very high winds to prevent damage. In small-scale generators, wind turbines start reducing speed in the 25 to 30 mile-per-hour range.

Friction, Turbulence, and Smart Siting

Now that you understand how winds are created, it's time to fine-tune your knowledge of wind. I'll begin by looking at a major force in the life of wind. It's called ground drag, and it occurs when air flows across a surface. Ground drag is a slowing of the wind closest to the ground as it flows across a surface. It is caused by friction.

Friction is the force that resists movement of one material against another. You create friction, for example, when you rub your hands together. (Try it.) Wind flowing across land or water, creates friction like rubbing a piece of wood on sand paper.

Ground drag due to friction varies considerably, depending on the texture or roughness of the surface. The rougher or more irregular the surface, the greater the friction. As a result, air flowing across the surface of a lake generates less friction than air flowing over a meadow; and air flowing over a meadow generates less friction than air flowing over a forest.

Ground drag slows air movement and extends to a height of about 1,650 feet (500 meters) above the Earth's surface. However, the greatest effects are closest to the ground—the first 60 feet (18 meters) above the ground over a relatively flat, smooth surface. Over trees, the greatest effects occur within the first 60 feet (18 meters) above the tree line.

A 20-mile-per-hour wind measured at 1,000 feet above the surface of the ground covered with grasses flows at 5 miles per hour at 10 feet. It then increases progressively until it breaks loose from the influence of friction. Figure 2.6 shows the difference in wind speed from 165 feet (50 meters) to 16.5 feet (5 meters). Figure 2.7 compares wind speed over a grassy area to wind speed over a forest, a significantly rougher surface. Notice that the wind speed increases more rapidly above the forest.

Understanding friction is important because of the dramatic way it influences wind speed near the surface of the ground where residential wind generators are, by necessity, located. Because the effects of friction decrease with height above the surface of the Earth, savvy installers typically mount their wind turbines on towers 80 to 120 feet high (24 to 37 meters), sometimes as high as 180 feet (55 meters) in forested regions, so their turbines are out of the influence of ground drag. At these heights, the

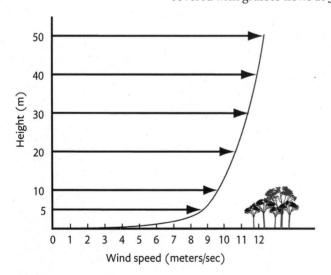

FIGURE 2.6. Effect of Ground Drag. Winds move more slowly at ground level due to friction; wind speed increases with height above ground, due to a reduction in friction. Credit: Anil Rao.

winds are substantially stronger than they are near the ground. As I will make clear shortly, a small increase in wind speed results in a substantial increase in the amount of power that's available from the wind and the amount of electricity a wind generator can produce. Mounting a wind turbine on a tall tower therefore maximizes the electrical output of the turbine. Placing a turbine on a short tower has just the opposite effect. It places the generator in the weaker winds and is a bit like mounting solar modules in the shade.

Ground drag isn't the only natural phenomenon that affects the production of wind turbines. Another is turbulence.

Turbulence results from air flow across rough surfaces. Objects in the way of the wind, such as trees or buildings, interrupt the wind's smooth laminar flow, causing it to tumble and swirl (Figure 2.8). Rapid changes in wind

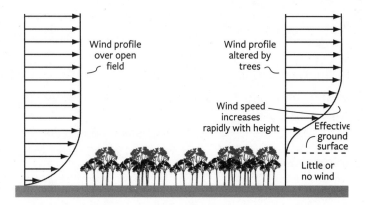

FIGURE 2.7. Wind Speed vs. Height. This graphic compares the wind speed over a grassy area and a forest. As you can see, a forest virtually eliminates ground-level winds, causing the effective ground level to shift upward. Note also that the wind speed increases more rapidly with height over a forest than over a grassy area. Wind turbines placed well above the treeline can avail themselves of powerful winds. Credit: Anil Rao.

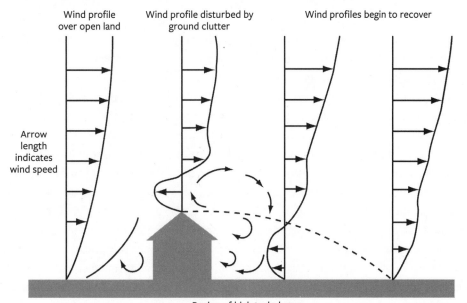

FIGURE 2.8. Ground Clutter and Turbulence. Trees, houses, barns, silos, billboards, garages, and other structures—collectively referred to as ground clutter—create considerable turbulence. Like eddies behind rocks in streams, the turbulent zone contains a fluid (air) that swirls and tumbles, moving in many directions. Turbulence reduces the harnessable energy of the wind and causes considerable wear and tear on wind machines that can shorten their lifespan. Credit: Anil Rao.

Turbulence is to a wind turbine like potholes to your car.

Robert Preus,
formerly with Abundant
Renewable Energy

speed occur behind large obstacles. In such instances, winds may even flow in the opposite direction to the general airflow. This highly disorganized wind flow is referred to as *turbulence*.

Turbulent wind flows wreak havoc on wind turbines, especially the less expensive lighter-weight wind turbines often installed on short towers by cost-conscious homeowners in an effort to save money. Buffeted by turbulent winds, wind turbines hunt around on the top of their towers, constantly seeking the strongest wind, starting and stopping repeatedly. This decreases the amount of electricity a wind turbine generates.

Turbulence not only decreases the harvestable power available to a wind generator, it also causes vibration and unequal loading (unequal forces) on the wind turbine, especially the blades, that may weaken and damage the turbine. Turbulent winds created by obstacles in the path of the wind therefore increase wear and tear on wind generators. Over time, they can destroy a turbine. The cheaper the turbine, the more quickly it will be destroyed, if placed in a turbulent location. A homeowner may find that a turbine he or she had hoped would produce electricity for 10 to 20 years only lasted two years. In extreme cases, an inexpensive, lightweight turbine can be destroyed in a matter of months.

When considering a location to mount a wind turbine, be sure to avoid turbulence-generating obstacles such as silos, trees, barns, houses, and even other wind turbines, a topic discussed in more detail in Chapter 4. Proper location is the key to avoiding the effects of turbulence. Turbulence can also be dramatically minimized by mounting a wind turbine on a tall tower—out of turbulence. Mounting a wind turbine on a tall tower offers four distinct benefits: (1) it situates the wind generator in the stronger, higher-energy-yielding winds, substantially increasing electrical production; (2) it distances the turbine from the damaging effects of turbulence; (3) it decreases the wind turbine's maintenance and repair requirements; and (4) it increases the wind turbine's useful lifespan substantially, perhaps tenfold. Longer turbine life means less overall expense—and more electricity from your investment.

As shown in Figure 2.9, all obstacles create a downstream zone of turbulent air, or turbulence bubble. As illustrated, the turbulence bubble typically extends vertically about twice the height of the obstruction. It also extends downwind approximately 15 to 20 times the height of the obstruction. For example, a 20-foot-high house creates a turbulence bubble that extends 40 feet above the ground and 300 to 400 feet downwind from it. As illustrated in Figure 2.9, the turbulence bubble also extends upwind a bit—about two times the object's height. In this case, the upwind bubble extends about 40 feet upwind from the house. The upstream portion of the

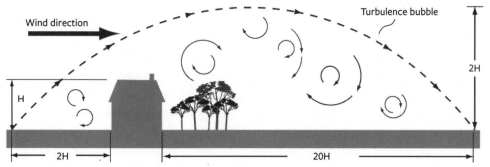

Wind direction

Turbulence bubble

2H

H

2H

20H

FIGURE 2.9. Turbulence Bubble. The turbulence bubble created by an obstacle or several obstacles like houses and surrounding trees, extends upwind, downwind, and even above the clutter. Credit: Anil Rao.

bubble is created by wind backing up as it strikes the obstacle—much like water flowing against a rock in a river or against the bow of a boat.

If 50-foot trees surrounded the house, the upwind bubble would extend about 100 feet. The bubble would extend downwind about 750 to 1,000 feet and vertically about 100 feet above the ground. Remember also that the bubble shifts as wind direction shifts. If the wind is blowing out of the north, the bubble extends to the south. If the wind is blowing from the south, the bubble extends to the north.

To avoid costly mistakes, installers recommend that wind turbines be mounted so that the complete rotor (the hub and the blades) of the wind generator is *at least* 30 feet (9 meters) above the closest obstacle within 500 feet (about 150 meters), or tree line in the area, whichever is higher (Figure 2.10). This should place a wind turbine out of the turbulence bubble into the stronger, smooth-flowing winds. Don't listen to those who recommend lesser heights. It's advice that you'll regret forever, as many unhappy customers have.

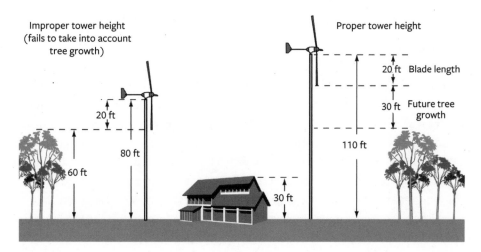

Improper tower height (fails to take into account tree growth)

Proper tower height

20 ft Blade length

30 ft Future tree growth

20 ft

80 ft

110 ft

60 ft

30 ft

FIGURE 2.10. Treelines and Wind Speed. Siting a wind turbine in an open area surrounded by trees is possible, but requires special attention to tower height. In this case, the top of the treeline is the effective ground level. The turbine needs to be mounted well above the trees to achieve optimum output. Credit: Anil Rao.

Raising a wind turbine into the smoothest, strongest winds ensures greater electrical production and longer turbine life—both money-making benefits of smart siting. If your home or business is in an open field surrounded by trees, as is often the case in some Midwestern and Eastern states, remember that trees can dramatically reduce the speed of the wind flowing across your property. The wind turbine needs to be above the tree line to be effective. Remember, too, when determining the appropriate tower height: trees grow and towers don't. Be sure to account for growth of nearby trees over the 20- to 30-year life span of your wind system when determining tower height. (I'll talk more about siting a wind turbine in Chapter 4.)

The Mathematics of Wind Power

I've been emphatic about the importance of placing a wind turbine on a high tower above the slower and more turbulent ground winds and will remind you of this essential law of wind energy production a few more times in this book.

To understand how important tall towers are, you need to know a simple mathematical equation. It allows us to calculate the power available to a wind turbine at different tower heights because of differences in wind speeds. This equation takes into account the three main factors that influence the output of a wind energy system: (1) air density, (2) swept area, and (3) wind speed.

The power available to a wind turbine is expressed in the equation:

$$P = \tfrac{1}{2}\, d \times A \times V^3$$

In this equation, P stands for the power available in the wind (not the power a wind generator will be able to extract—that's influenced by efficiency and other factors). Density of the air is d. Swept area is A. Wind speed is V. Of the three factors that influence power (P), air density is the least important.

Air Density

Air density is the weight of air per unit volume. Air density varies with elevation, but doesn't have much of an effect on power until one reaches 2,500 feet (about 760 meters) above sea level. As shown in Table 2.1, at 3,000 feet (about 910 meters) above sea level, the air density is only 9 percent lower than at sea level. At 5,000 feet (about 1,525 meters) above sea level, air density has dropped by about 15 percent. At 8,000 feet (about 2,440 meters) above sea level, air density plummets by about 24 percent. You can anticipate a decrease in the expected output of a wind turbine of about 3 percent per 1,000-foot (about 300 meters) increase in elevation.

In general, then, for higher elevation areas like my former home in the foothills of the Rockies, at 8,000 feet above sea level, the air is considerably less dense than lower-elevation air at my home in Missouri, at 800 feet above sea level. A wind turbine in a 12-mile-per-hour wind in Missouri will produce more electricity than an identical wind turbine at my former residence in Colorado, assuming they are both mounted at the same height and are free from obstructions. For example, if I installed a Bergey Excel 6, a wind turbine produced by a pioneer in small wind energy, Mike Bergey, in both Colorado and Missouri, and the turbines were operating in a 12-mile-per-hour average wind, the Missouri turbine would produce about 11,387 kilowatt-hours of electricity per year. (See sidebar for a definition of kilowatt-hours.) The same turbine at 8,000 feet above sea level would produce 8,867 kilowatt-hours per year, based on the manufacturer's estimate of energy production.

Air density is also a function of humidity—the amount of moisture contained in air—and air temperature. Consider humidity first. Although humid air seems more dense to us than dry air, it's not. It is actually less dense than drier air at the same temperature.[1] As a result, a wind turbine in a 14-mile-per-hour wind in arid New Mexico will produce slightly more electricity than would the same generator in humid Florida or Missouri. (The difference, however, is negligible.)

Air density is also affected by temperature. As you would expect from previous material in this chapter, warmer air is less dense than colder air. Consequently, a wind turbine operating in cold (denser) winter winds blowing at 12 miles per hour would produce slightly more electricity than the same wind turbine in warmer winds blowing at the same speed.

Although temperature and humidity affect air density, they are not factors we can change. Homeowners and installers do need to be aware of the reduced energy available at higher altitudes, however, so they don't create unrealistic expectations for wind-electric systems installed in such locations.

Although density is not a factor we can control, wind installers do have control over a couple of other key factors, notably, swept area ("A," in the equation

Table 2.1. Air Density vs. Wind Turbine Performance

Elevation (ft)	Elevation (m)	Air Density
0 (sea level)	0 (sea level)	100%
500	152	99%
1,000	305	97%
2,000	610	94%
3,000	915	91%
4,000	1,220	88%
5,000	1,524	85%
6,000	1,829	82%
7,000	2,134	79%
8,000	2,439	76%
9,000	2,744	73%
10,000	3,049	70%

Kilowatt-hours

Energy produced by a wind turbine and consumed in by electrical devices is measured in kilowatt-hours, or kWh for short. Kilowatt-hours is determined by multiplying the watts a device produces or consumes by time in hours. A device that uses 1,000 watts for one hour, for example, produces 1,000 watt-hours or 1 kilowatt-hour.

given just above) and, believe it or not, wind speed. As you shall soon see, both of these factors have a much greater impact on the amount of power available in wind and the electrical output of the turbine than air density does (which is, as just noted, affected by elevation, humidity, and temperature).

Swept Area

Swept area, a topic I'll discuss in greater depth in Chapter 5, is the area of the circle the blades of a wind turbine create when spinning (Figure 2.11). I like to think of it as a wind turbine's collector surface. Just like a solar array, the larger the collector surface, the more power a wind turbine can capture from the wind.

Swept area is determined by blade length.[2] The relationship is pretty straightforward: The longer the blades, the greater the swept area. The greater the swept area, the greater the electrical output of a turbine.

As the equation suggests, the relationship between swept area and power output is linear. Theoretically, a ten percent increase in swept area will result in a ten percent increase in electrical production. Doubling the swept area doubles the output. When shopping for a wind turbine, bear in mind that relatively small increases in blade length will result in very large increases in the swept area. That's because the area of the circle spinning blades "describe"— the swept area—is a function of the radius (blade length) squared. For the mathematically inclined, that's: $A = \pi \times r^2$.

In this equation, the funny-looking Greek symbol is pi, which is a constant: 3.14. The letter r stands for the radius of a circle, the distance from the center of the circle to its outer edge. For a wind turbine, radius is the same as the length of a wind turbine's blade.

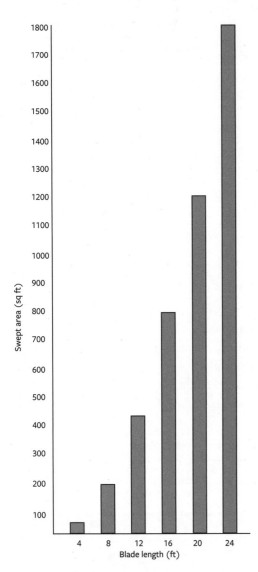

FIGURE 2.11. Blade Length and Swept Area. A small increase in radius or blade length results in a large increase in swept area. Credit: Forrest Chiras.

Some Important Conversion Factors to Know

To convert meters per second to miles per hour, simply multiply meters per second by 2.24. In other words, 1 m/s = 2.24 miles per hour. To convert miles per hour to meters per second, multiply mph by 0.45. In other words, 1 mph = 0.45 m/s.

When shopping for a wind turbine, always convert blade length to swept area, if the manufacturer has not done so for you (they usually do). Shop for wind turbines with the longest blades and hence largest swept area to ensure maximum energy production.

Figure 2.11 shows the relationship between blade length and swept area. Notice how small increases in blade length result in large increases in swept area. Here's an example to help drive home this important point: a wind generator with an 8-foot blade has a swept area of 200 square feet. A wind generator with a 10-foot blade—that is, a 25 percent longer blade—has a 314 square-foot swept area. In this case, a 25 percent increase in blade length results in a 57 percent increase in swept area and, theoretically, a 57 percent increase in electrical production.

Wind Speed

Although swept area is more important than the density of air, wind speed is even more important when it comes to the output of a wind turbine. That's because the power available from the wind increases with the cube of wind speed. This relationship is expressed in the power equation as V^3 ($V \times V \times V$), or wind speed multiplied by itself three times. Let's consider a simple example provided by Mick.

Suppose that you mount a wind turbine (foolishly, we might add), at 18 feet (5.5 meters) above the ground surface on the grass-covered plains of Kansas. Let's suppose that you measure the output when the wind is blowing at eight miles per hour (3.6 meters per second). A savvy friend, who knows how important it is to mount a wind turbine on a tall tower, installs an identical wind turbine on a 90-foot (27 meter) tower. He also measures the output. When the wind is blowing at 8 miles per hour at 18 feet where your wind turbine flies, it is blowing at 10 miles per hour at 90 feet, the height of your neighbor's turbine. Wind speed is 25 percent higher.

In this example, the power available to both turbines can be approximated by multiplying the wind speeds by themselves three times. (Units aren't important for this comparison.) For the lower turbine, the $8 \times 8 \times 8 = 512$. The power available to the wind turbine mounted on a 90-foot tower is 10 cubed, or $10 \times 10 \times 10$, or 1,000.

This example shows that a two-mile-per-hour increase in wind speed—only a 25 percent increase—doubles the available power. Put another way, a 25 percent increase in wind speed yields an increase of nearly 100 percent.

Although winds are out of the control of us mortals, we can affect the wind speed to which our wind turbines are exposed by choosing the best possible site on our property and by installing our wind generators on the tallest towers. Remember: the important lesson of V^3 is that a small increase in wind speed results in a very large increase in the power available to a wind turbine and the electrical output of the unit.

Remember for future discussions with those who seem to miss this essential point: a 10 percent increase in wind speed will always result in a 33 percent increase in power available in the wind. That's the reason why I'll remind you repeatedly throughout this book about the importance of proper siting—situating wind turbines out of the turbulence bubble and mounting them as high as possible.

Here's an example that will make the point one more time: I installed a Skystream 3.7 on a 126-foot tower on our property high above the trees. A resident of the same county installed the same turbine on a 35-foot tower. Her turbine produces 250 kWh of electricity per year. Mine produces ten times that amount.

These examples show why you should always be suspicious of anyone who tells you that it's fine to mount a wind turbine on a short tower in your backyard. It's the worst possible advice. I've heard it given by professional installers and a certain wind turbine manufacturer (now out of business) who seemed more interested in selling wind turbines than the performance of those turbines and customer satisfaction.

Closing Thoughts

Wind is a tremendous resource available in many parts of the world thanks to the Sun, or, more specifically, the Sun's unequal heating of the Earth's surface. Depending on your location, you may be able to take advantage of offshore and onshore winds or perhaps mountain-valley winds. More likely, prevailing winds and storm winds will become your greatest ally in achieving energy independence and reducing your carbon footprint. Just don't forget to take into account ground drag created by friction that could slow wind speed near your site and turbulence that can rob you of additional energy and tear your beloved wind turbine to pieces. Site wisely on a tall tower and you'll be repaid day after day after day.

CHAPTER 3

Wind Energy Systems

One of the first decisions you will need to make—after determining whether your site is appropriate for wind—is the type of system you need to meet your needs. Wind-electric systems fall into three categories: (1) grid-connected, (2) grid-connected with battery backup, and (3) off-grid. In this chapter, I'll examine each system. I'll describe their components and discuss the pros and cons of each one. I'll also examine ways to couple wind systems with other renewable energy technologies such as solar electricity to create a hybrid system. This information will help you decide which system suits your needs and lifestyle. Before we examine the types of wind energy systems, let's take a brief look at their main components.

Components of a Wind Energy System

Wind systems contain many components. To begin, though, I'll focus our attention on two of them: the wind turbine and the tower. Subsequent chapters contain more detailed discussions of these and other components.

Wind Turbines

The wind turbine or wind generator is a key component of all wind-electric systems. Although there are several types of wind generators, most wind turbines in use today are horizontal-axis units, or HAWTs (this term is explained shortly). Most small wind turbines in use today have three blades attached to a central hub. Together the blades and the hub are called the rotor. The rotor, in turn, is often connected directly (or, less commonly, via a shaft) to an electrical generator. In turbines with a shaft, the shaft runs horizontal to the ground, hence the name, horizontal-axis wind turbine. When the rotor turns, the generator produces alternating current

AC vs. DC Electricity

Electricity comes in two basic forms: direct current (DC) and alternating current (AC). Direct current electricity consists of electrons that flow only in one direction through the electrical circuit; it's the kind of electricity produced by flashlight batteries or the batteries in cell phones, laptop computers, or portable devices such as iPods. It is also the kind of electricity produced by photovoltaic modules.

Most wind turbines produce alternating current electricity. In alternating current, the electrons flow back and forth. That is, they switch or alternate direction in very rapid succession, hence the name "alternating current." When electrons flow from left to right and then back again, we refer to this as a cycle.

In North America, electric utilities produce electricity that cycles back and forth 60 times per second. It's referred to as 60-cycle-per-second— or 60 hertz (Hz)—AC. The unit *hertz* commemorates Heinrich Hertz, the German physicist whose research on electromagnetic radiation served as a foundation for radio, television, and wireless transmission. In Europe and Asia, the utilities produce 50-cycle-per-second AC.

Alternating current electricity is produced by most modern wind turbines; when used to charge batteries, though, it is often converted to DC electricity. AC electricity is also produced by the generators in hydroelectric and power plants that run on fossil fuels, like coal, or nuclear fuels. No matter what form of energy is used to turn a generator, all of them operate on the principle of magnetic induction—they move a wire through a magnetic field (or vice versa) to produce electricity. This is true for both AC and DC generators.

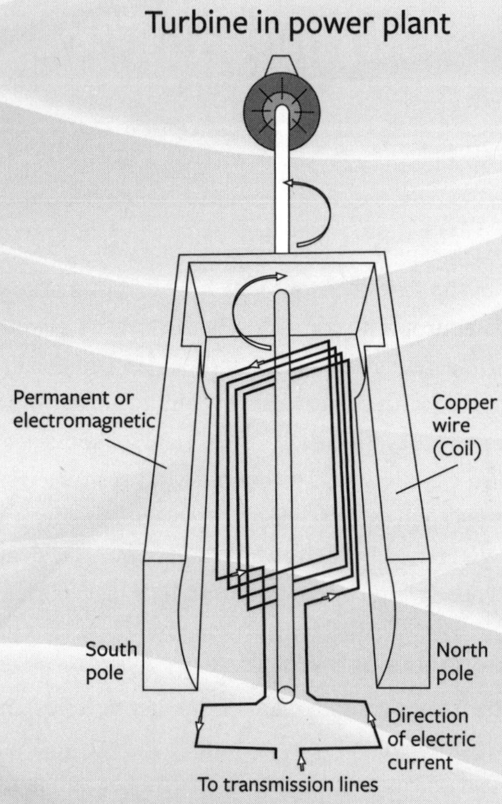

FIGURE 3.1. Electromagnetic Induction. Moving a conductor like a copper back and forth through a magnetic field generates alternating current electricity. Moving a magnet back and forth over a copper wire has the same effect. In most wind turbines, magnets are spun by stationary coils of copper wires.
Credit: Forrest Chiras

(AC) electricity. (See the accompanying textbox for an explanation of AC electricity.)

One of the key components of a successful wind generator is the blades. They capture the wind's kinetic energy (air in motion) and convert it into mechanical energy (rotation) that is converted into electrical energy by the generator. I'll discuss their design, how they work, and the materials they are made of in Chapter 5.

The generators of wind turbines are often protected from the elements by a durable housing made from fiberglass or aluminum (Figure 3.2). However, in some modern small wind turbines, the generators are exposed to the elements. Interestingly, an alternator can deal with the weather and often benefits from the natural cooling that occurs when it is exposed directly to the air.

Wind generators produce electricity when the wind blows—but only above a certain minimum speed. This is known as *the cut-in speed*.

Most wind turbines in use today have tails (aka tail vanes) that keep them pointed into the wind to ensure maximum production. However, some turbines, like the Skystream 3.7 now manufactured by Xzeres in Oregon, have no tails. They are designed to orient themselves to the wind without them. (More on these turbines in Chapter 5.)

FIGURE 3.2. Wind Turbine Design. The generators in most small wind turbines are often housed in a protective case made from aluminum or fiberglass. Credit: Bergey WindPower.

Towers

Much like the star quarterback on a football team, the wind turbine typically receives most of the attention in discussions on small wind energy. However, like football, wind energy is a team effort. Several other components are essential parts of the team. They play key roles in capturing and converting the wind's energy, roles that help the star of the team, the wind turbine, shine.

One vital team member is the tower. Although I'll take a much closer look at towers in Chapter 6, a few words are in order now. To begin, residential wind generator towers fall into three broad categories: (1) freestanding, (2) guyed, and (3) tilt-up (Figure 3.3).

Freestanding towers may be either monopoles or lattice structures. Freestanding monopole towers consist of high-strength hollow tubular steel like those that

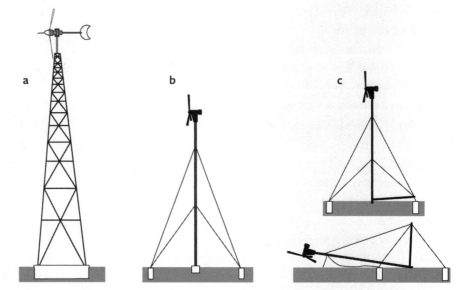

FIGURE 3.3. Tower Types. Three types of towers in use today: (*a*) freestanding, (*b*) fixed, guyed, and (*c*) tilt-up. Note that both freestanding and fixed, guyed towers may be monopoles or lattice towers. Tilt-up towers are typically monopole towers. Credit: Anil Rao.

support streetlights. Lattice towers consist of tubular steel pipe or flat-metal steel that is bolted or welded together to form a lattice structure like the famed Eiffel Tower in Paris.

To be successful, freestanding towers must be strong enough to support the weight of the wind turbine, but more importantly, they need to be strong enough to withstand the forces of the wind acting on the turbine. (Any tower strong enough to do this is strong enough to support the weight of the turbine.) Towers must also have a large foundation to counteract the tremendous forces applied by the wind, forces that could easily topple a weak or poorly anchored tower. Large amounts of steel and concrete are required to accomplish this task. This makes freestanding towers the most expensive tower option.

The second tower type, the guyed tower, is shown in Figure 3.3b. Guyed towers may be made of high-strength steel pipe, or they may be lattice structures. They are supported by high-strength woven steel cables, known as *guy cables*. The guy cables extend from tower attachment points to concrete or steel anchors attached securely to the ground. Guy cables stiffen towers so they don't buckle, and they hold them in place so they don't lean or fall over in strong winds.

Because guy cables strengthen and stabilize the tower, less steel is required for a guyed tower than a freestanding tower.[1] Because of this, they are also considerably less expensive than freestanding towers. As illustrated in Figure 3.3b, guy cables are attached to the tower every 20 to 30 feet, although wider spacing is also possible. A 120-foot-tall tower may require four sets of cables.

The third type of tower, and one that is very popular, is the tilt-up tower, so named because it can be raised and lowered—tilted up and down. This feature makes it possible to inspect, maintain, and repair a wind turbine from the ground. Tilt-up towers typically consist of high-strength steel pipe or a lattice structure supported by guy cables.

When it comes to the performance of a wind system, towers are as important as the wind turbine itself. Unfortunately, many wind turbines are mounted on towers that are too short. This occurs out of ignorance and misguided frugality. As I will point out numerous times in this book, installing too short a tower is a foolish, potentially costly mistake. Mounting a wind generator on a short tower is akin to mounting a solar-electric module in the shade. As I pointed out in the previous chapter, properly sized towers place wind turbines in the stronger, smoother, more powerful winds. Properly sized towers also raise turbines out of turbulence created by ground clutter that severely diminishes the quality and quantity of wind. Returning to our football analogy, towers are like the offensive line in a football team. They protect the star player, the turbine, making its life less rough and tumble and ensuring much higher productivity. Don't let anyone talk you into a short tower!

Wind systems also require wires that run down the tower, transporting electricity produced by the turbine to the point of use. These wires attach to leads on the generator and may run externally (for example, alongside a tower leg of a lattice tower) or internally (inside a tubular tower) to a junction box at the base of the tower. From there, they typically run underground in conduit to the house, barn, shop, or business where the remaining system components are housed.

As you shall soon see, wind-electric systems involve a number of additional components. Most wind systems require a controller, an inverter, and an AC service panel (the breaker box through which electricity from the wind system flows to household circuits). Battery-based wind energy systems require batteries to store surplus electricity and a device known as a *charge controller* that helps protect the batteries. Wind systems, like other renewable energy systems, also require disconnects, meters, and protective devices such as fuses, circuit breakers, and lightning arresters. I'll describe the function of each of these components as we explore the three main types of wind energy systems. More detailed discussions of key components, like towers, turbines, inverters, and batteries are found in subsequent chapters.

Wind Energy System Options

As noted in the Introduction, wind systems fall into three main categories: (1) grid-connected, (2) grid-connected with batteries, and (3) off-grid. Each one is designed for a specific purpose. You'll need to be very clear up front which one will meet your needs. This section will help you figure out which type of system you need. Let's begin with grid-connected systems.

Grid-connected Systems

The simplest of the major wind energy systems is the grid-connected option, shown in Figure 3.4. Grid-connected systems are so named because they are connected directly to the electrical grid—the vast network of electric wires that crisscross cities, towns, states, and nations. They're also referred to as *batteryless grid-connected* or *batteryless utility-tied* systems because they do not employ batteries to store surplus electricity.[2]

FIGURE 3.4. Grid-connected Wind System. The grid-connected wind system is the simplest of all systems. Wild or variable AC electricity produced by the turbine is first fed into the controller. Here it is converted to DC electricity. The DC electricity is fed into the inverter, which produces grid-compatible AC electricity to power household loads. Surpluses are backfed onto the grid. Credit: Forrest Chiras.

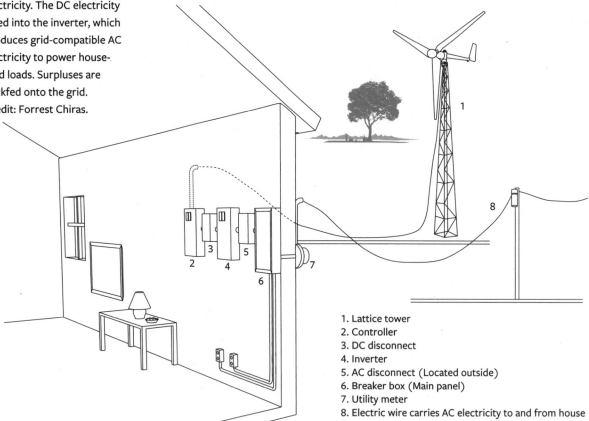

1. Lattice tower
2. Controller
3. DC disconnect
4. Inverter
5. AC disconnect (Located outside)
6. Breaker box (Main panel)
7. Utility meter
8. Electric wire carries AC electricity to and from house

In batteryless grid-tied systems, the electrical grid accepts surplus electricity—electricity produced by the turbine in excess of household demand. When a wind system is inactive, the grid supplies electricity to the home or business. In a sense, then, the grid serves as the storage medium.

As shown in Figure 3.4, a batteryless grid-connected system consists of nine main components: (1) a wind generator, specifically designed for grid connection, (2) a tower, (3) a power conditioner (labeled DC control panel), (4) a DC disconnect, (5) an inverter, (6) a utility disconnect (labeled AC disconnect), (7) the main service panel, (8) a utility meter, and (9) the electrical grid.

To understand how a batteryless grid-connected system works, let's begin with the wind generator. In most batteryless grid-connected systems, the wind generator produces "wild AC" electricity.[3] Wild AC is alternating current with a frequency (cycles per second) and voltage that vary with wind speed. Wild AC is also referred to as *variable AC*, a much more accurate term. (See sidebar for a definition of frequency and voltage.) In most small wind turbines, the frequency and voltage are very low when the blades begin to spin. The faster the blades spin, the higher the voltage and the greater the frequency.

Variable AC, produced by wind turbines, is not directly usable. Appliances and electronic devices in our homes require a tamer version of electricity—alternating current with a relatively constant frequency and voltage, like that available from the grid. In a grid-connected system, then, the wild voltage must first be "tamed." That is, its frequency and voltage must be converted to the same frequency and voltage that the utility provides—usually 240 volt, 60 cycle per second electricity. This is fed into the breaker box or main service panel from which both 120- and 240-volt circuits are wired. Most circuits in a home—like those that power refrigerators, lights, and outlets—are wired at 120 volts. Two hundred and forty volt circuits are reserved for the heavy hitters like electric water heaters, electric dryers, electric stoves, electric furnaces, geothermal systems, and minisplits (air source heat pumps).

Variable AC is converted to standard household AC in two steps. First, the controller in the system converts (rectifies) the wild AC to DC. This function is performed by an electronic device called a rectifier,

Frequency and Voltage

Alternating current electricity is characterized by a number of parameters. Two of them are frequency and voltage. Frequency, described in the text, refers to the number of times electrons switch direction every second and is measured as cycles per second. (One cycle is a switch from flowing from the left to the right and then back again.)

The flow of electrons through an electrical wire is created by an electromotive force that scientists call voltage. The unit of measurement for voltage is volts. You can think of voltage as electrical pressure, or the driving force that causes electrons to move. The higher the voltage, the more pressure.

which is typically located in a controller, shown in Figure 3.4. (In some systems, the "controller" rectifiers are in the inverter.) DC electricity flows from the controller to the next component of the system, the inverter, also shown in Figure 3.4. The inverter converts the DC electricity to grid-compatible AC—240-volt, 60 cycle per second electricity. Because the inverter produces electricity in sync with the grid, it's often referred to as a synchronous inverter. It's also known as a *utility interactive inverter* or *grid-tied inverter*.

While grid-compatible wind generators typically produce variable AC, another type of wind generator has been introduced into the small wind market. It is known as an *induction generator*. Explained in more detail in Chapter 5, an induction generator is similar to an electric motor that, when driven fast enough by the spinning blades of the wind turbine, produces grid-compatible AC electricity. (This is the way large commercial wind turbines on wind farms work.) As a result, its output does not need to be converted to DC and then inverted back to AC, eliminating a few of the costly components in a small wind energy system.

The 240-volt AC produced by the inverter (or directly by an induction generator) flows to the main service panel. From there, it flows to active loads—that is, to electrical devices drawing power. If the wind turbine is producing more electricity than is needed, which is often the case; the excess automatically flows—is backfed—onto the grid.

As shown in Figure 3.4, surplus electricity backfed onto the grid travels from the main service panel through the utility's electric meter, typically mounted on the outside of the house. It then flows through the wires that connect the home or business to the grid. The surplus electricity then travels along the power lines where it flows into neighboring homes or businesses. The utility treats the electricity as if it were theirs, and the end users pay the utility directly for the electricity you provided to them. Fortunately, the utility meter monitors a home's or business's contribution to the grid at all times so the utility can credit the local producer for his or her contribution. The meter also keeps track of electricity the power company supplies to the home or business when the wind system is not generating. To learn how the electric company measures what you are putting onto the grid and how they "pay" for it, check out the accompanying textbox, "The Ins and Outs of Net Metering."

In addition to the electric meter—or meters (some utilities install two additional meters)—that monitor the flow of electricity onto and off the local utility grid, Code-compliant grid-connected wind energy systems also contain safety disconnects. These are manually operated switches that enable service personnel to disconnect at a couple of key points in the system to prevent electrical shock if service is required. As shown in Figure 3.4, a DC disconnect is wired into the system between the controller and the inverter. An AC disconnect is wired between the inverter and main service panel. When switched off, it disconnects and isolates the wind energy system from the household circuits and the grid. The AC disconnect must be mounted outside so that it is accessible to utility company personnel (Figure 3.4). That way, utility workers can isolate the wind system from the grid if they need to work on the electric lines in your area without fear of shock—for instance, if an electric line goes down in an ice storm. The AC disconnect must also contain a switch that can be locked in the off position by the utility worker so that the homeowner or a family member doesn't accidentally turn the system back on prior to their completion of repairs (Figure 3.5b). For some time, the lockable disconnect was considered critical for the safety of utility personnel, because the 240-volt electricity from an inverter becomes thousands of volts on the utility distribution line after it passes through the "service-drop" transformer.

Interestingly, lockable AC disconnects are not required by all utilities. Moreover, the large Colorado and California utilities, with thousands of solar- and wind-electric systems, have dropped the requirements for utility company-accessible, lockable disconnects. They've found that they haven't got the personnel to lock out all the renewable energy systems, and experience has shown that they don't have to. As discussed shortly, grid-compatible inverters automatically shut off when the utility power goes down. Properly installed wind-electric systems will not backfeed onto a dead grid, with or without a disconnect. Period.

As I'll discuss in Chapter 7, grid-compatible inverters monitor line voltage and frequency (the frequency and voltage of electricity on the grid). When they detect a change in either, for instance, a drop in voltage due to a power outage, the inverter automatically shuts down—and it stays off until the situation is corrected. As a result, no electricity can flow onto the grid. Although this automatic disconnect feature works reliably, when

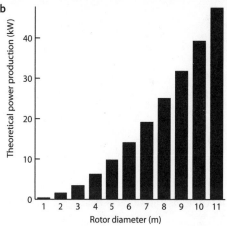

FIGURE 3.5. Utility Disconnects. Most utilities require (*a*) a utility disconnect located near the main service entrance of a home or business. (*b*) This disconnect must be accessible, lockable, and visible, that is, contain a knife-blade switch that can be visually inspected to be sure that there's no electrical connection. Credit: Dan Chiras.

it comes to safety, a second line of defense is prudent. The second line of defense is the lockable disconnect switch. "Lockable disconnects are theoretically unnecessary, because UL-listed inverters meet interconnection safety standards and can be relied upon to disconnect if the grid goes down," notes electrical engineer Robert Aram. "But when human life is at stake, many believe that prudence dictates that you not rely on only one system for safety.... The lockable disconnect is a backup to the safety features of the inverter. It is also on the outside of the building, where the utility worker can see it—and see that it is open (disconnected). Utility workers can also put their personal locks on them to be sure that it stays open (disconnected).... The inverter is inside the building where the utility worker cannot see it. How does he know that you have a UL-listed inverter? Perhaps you have kluged something together that feeds directly into the grid with no safety functions?...Every source of power in a utility's system, every generator and every transformer, has a visible disconnect switch. Since the utilities have backup safety disconnects for all their own equipment, isn't it reasonable for them to require renewable energy system owners to have one also?"

My response to this is that applications to install a grid-tied system should be reviewed by all utilities prior to installation. The applications and drawings submitted with them should stipulate the installation of UL-listed inverters that shut down when they sense a power outage. Utilities usually inspect systems prior to allowing their owners to connect to the grid. At that point, they should verify the presence of a UL-listed inverter and test it to be sure it shuts down if the grid is shut down. (The specific listing that ensures an inverter will shut down automatically is UL-1741.)

The Pros and Cons of Grid-connected Systems

Batteryless grid-connected systems represent the majority of all new wind systems in the United States. Although popular, they do have their pluses and minuses, summarized in Table 3.1.

On the positive side, batteryless grid-connected systems are relatively simple. And, as such, they are generally less expensive than other options. They are often 25 percent cheaper than battery-based systems.

Because batteryless grid-connected systems contain no batteries, they require less maintenance. Operators won't have to manage and maintain a costly battery bank, a topic discussed in Chapter 8.

Table 3.1. Pros and Cons of Batteryless Grid-tie Systems

Pros	Cons
Simpler than other systems	Vulnerable to grid failure unless an uninterruptible power supply (generator) is installed
Less expensive	
Less maintenance	
More efficient than battery-based systems	
Unlimited storage of surplus electricity	
Greener than battery-based systems	

The Ins and Outs of Net Metering

The idea of selling electricity to a local utility appeals to many people. In most new installations, utility companies install digital meters that tally electricity delivered to and supplied by a home or business. These meters keep separate totals of the electricity coming from and going to the grid.

All customers who connect their wind systems to the grid enter into a contractual agreement—called an *interconnection agreement*—with their utility. It spells out many details, including the provisions for paying a customer for surplus. This payment language is part of a net metering policy established by the state. Nearly every state has one.

Net metering policies vary from state to state, but all provide rules regarding several key factors: (1) which types of renewable energy electricity-generating systems the rules pertain to (for example, wind-electric or solar-electric); (2) the maximum size a utility customer's system can be; (3) which utilities must abide by the rules (that is, municipal, investor-owned, or electric co-ops), and finally; (4) how customers are reimbursed for surpluses. Surplus electricity is referred to as *net excess generation* (NEG).

Two types of net metering policies exist: monthly and annual. First, though, let me point out something. The "net" in net metering refers to the fact that the customer is billed only for net consumption, i.e., what remains after credits for surpluses are applied. In states like Colorado and Kansas that offer annual net metering, net excess generation is reconciled once a year. This billing arrangement is known as annual net metering.

Annual net metering is a lot like many cell phone plans. In a cell phone plan, surpluses can be carried from one month to the next for a year. If you have a surplus of 500 minutes one month, they can be used, free of charge, the next month or the next—up to one full year.

In annual net metering, utilities carry surplus kilowatt-hours from one month to the next—up to a year. Because of this, surpluses generated in windy winter months can make up for shortages in the less-windy summer. For example, if your system generates a surplus of 2,000 kilowatt-hours in the winter, and you need an extra 1,000 kilowatt-hours in July and another 1,000 kilowatt-hours in August, the electricity is yours for free.

So, what about surpluses, if any, that exist at the end of the annual billing period?

In annual net metering, unused electricity remaining in the account at the end of the year is handled in one of three ways. They are: (1) transferred to the utility (forfeited); (2) purchased by the utility at the retail price of electricity, i.e., the same price that the customer pays the utility for electricity; or (3) purchased by the utility at its wholesale rate. Wholesale rate is referred to as *avoided cost*. It's what the electricity cost that utility—either to generate it or buy it from another supplier. Avoided cost settlement is less desirable than reimbursement at retail. Once the account is reconciled, the balance is set to zero, and the net metering starts over for the following year.

Some states (such as extremely renewable-energy-friendly Colorado) allow customers to carry their balance from year to year. Thus, if you have a surplus in 2017, you can carry it over to 2018. If you need it in 2019, it's yours for free.

The advantage of annual net metering is that it accommodates the seasonal variation in a wind system's production. In the late fall, winter, and early spring in many areas of the world, wind systems produce more electricity than is consumed by a customer. In the late spring, summer, and early fall, wind systems typically produces less electricity than is consumed. Surpluses can be withdrawn from the "bank" and applied to bills in those months.

Consider an example: Suppose that a customer with a wind system that delivered 1,000 kWh of electricity to the grid during the month of March, a fairly windy month in most locations in North America. Suppose also that the customer consumed 600 kWh from the grid during that month. For this month, the customer would be credited with the net production of 400 kWh and would not be billed for any electricity, although she would be billed for normal customer service and fees of $5 to $25. She would carry the 400-kWh surplus over to April. Let's assume that in April and May, the customer's surplus was 1,200 kWh of electricity. At this point, the customer has banked 1,600 kWh (the net consumption). Then June rolls around and the winds die down. Let's suppose the customer only generated 300 kWh of electricity from her wind system but consumed 800. In this month, she and her family had to draw 500 kWh of electricity from the grid. No worries. She's banked 1,600 kWh. Since the balance in the account was sufficient to cover the net consumption, the customer would only be billed the monthly service charge. In July, she can continue to draw off the net excess generation as well. If a surplus remains in August and September, the only fee the customer would pay would be the service fee. By October, as the winds begin to pick up, the customer may be out of surpluses, but could be producing enough to meet her needs.

In many states, utilities reconcile the customer's electric bills each month. This arrangement is known as monthly net metering. In monthly net metering, net excess generation is carried from one day to the next for a month. As a result, a surplus generated on Monday can be used on Friday. However, any surpluses remaining at the end of the month must be reconciled. They can't be carried over. To see how this works, consider the following example.

Suppose that the customer produced a surplus of 200 kWh of electricity in the first two weeks in the month of August. Because of heavy air conditioner use, however, the customer consumed 200 kWh more than the wind system produced during the remainder of month. In this case, the customer would not be charged for electricity. The first two-week surplus would offset the excess consumption in weeks three and four. The customer would only pay the service fee.

Now suppose that when September rolls around, it's particularly windy. Because of this, the customer produced a surplus of 500 kWh. That is to say, she produced 500 kWh of electricity more than she withdrew from the grid. Let's also suppose that the utility charges 15 cents per kilowatt-hour for electricity supplied by the grid and credits customers the same amount for surpluses delivered to the grid. If this system is in a state that offers monthly net metering and reimburses at retail rate what customers pay, the solar customer would receive a check or credit for $75 for the surplus. (That's 500 kWh net excess generation, at 15 cents.)

If you're thinking that this could be profitable venture, don't get your hopes up. There aren't many states that reimburse customers at retail rates for monthly net excess generation. Most utilities pay for surpluses at the avoided cost—the cost of generating power. In Missouri, for instance, its monthly net metering policy allows utilities to reimburse at avoided cost.

If you are paying the utility 10 to 12 cents per kilowatt-hour, they'll pay you about 2.5 cents. Some states, like Arkansas, simply "take" the surplus without payment to the customer—it all depends on state law. (If you're not happy with your state law, consider working to change it!)

Monthly net metering is generally the least desirable option, especially if surpluses are "donated" to the utility company or reimbursed at avoided cost. Annual reconciliation is a much better deal. It permits winter surpluses to be "banked" to offset summertime shortfalls. However, don't forget that even with annual net metering, the end-of-the-year surplus, if any, may be lost—forfeited to the utility, if state law permits this option.

The ideal arrangement, from a customer's standpoint, is a continual rollover. In such instances, there's no concern about losing your banked solar-powered kilowatt-hours. (Thirteen states permit continued rollover.)

Net metering is mandatory in many states, thanks to the hard work of renewable energy activists and forward-thinking legislators. At this writing (February, 2016), 46 states and the District of Columbia have implemented statewide net metering programs. Only three states have no net metering policy at this writing: Mississippi, Tennessee, and South Dakota. Even so, three utilities in these states have adopted net metering. Texas and Idaho have a voluntary program, which means the utilities can do whatever they want. Table 3.2 lists the states with the best policies. For a summary of net metering policy by state, see the Appendix. In Canada, some provinces have adopted net metering, too.

Although virtually all states have passed net metering policies, there are major differences. Differences include: (1) who is eligible, (2) which utilities are required to participate, (3) the size and types of systems that qualify, and (4) reimbursement for NEG.

Eligibility is usually fairly broad. States allow homeowners and businesses of all sorts to participate. Most states also require net metering for all types of utilities.

Where policies vary the most is the size of PV systems businesses and homeowners can install. The size of systems varies widely, from 10 kW to 2,000 kW. Illinois, for instance, places a limit of 40 kW on PV systems while California allows systems up to 1,000 kW. New Mexico permits systems up to 80,000 kW.

Some states also limit size based on a customer's typical annual energy consumption. This gets a bit tricky. Colorado, for instance, allows residential customers to install up to 10 kW and businesses to install systems up to 25 kW, but the projected output of the system legally cannot exceed 120% of a customer's annual electrical demand. If 120% of a customer's demand could be met by a 6-kW system, that's the largest system he or she can install. There are ways around this, however. If, for example, you are getting married and your spouse is bringing two children to the marriage, you could make a case for a system larger than what you needed when you were living alone.

Table 3.2. States with the Most Favorable Net Metering Policies

Arizona	New Jersey
California	New York
Colorado	Ohio
Connecticut	Oregon
Delaware	Pennsylvania
Maryland	Utah
Massachusetts	Vermont
New Hampshire	West Virginia

Source: "Best and Worst Practices in State Net Metering Policies and Interconnection Procedures," Freeingthegrid.org.

Nor will the owner have to replace costly batteries every five to ten years. Moreover, no batteries means no battery box or battery room, which may be costly to add to an existing home or incorporate in a new home. All of these factors add up to potentially huge savings.

Another substantial advantage of batteryless grid-connected systems is that they can store an unlimited amount of electricity (so long as the grid is operational). Although these systems don't store excess electricity on site like a battery-based system for later use—for example, during periods when the wind's not blowing—they "store" surplus electricity on the grid in the form of a credit on your utility bill. When winds fail to blow—or when a wind turbine isn't producing enough electricity to meet demands—electricity can be drawn from the grid, using up available credits from previous surpluses. The grid serves as an unlimited battery bank, meaning it can absorb as much surplus as a homeowner can generate. A battery bank is limited. When it is full, it's full. Surpluses generated in renewable energy systems not connected to the grid are typically dumped—that is, used to provide supplementary heat for living space or for domestic hot water so as not to be wasted. (For more on grid storage, see sidebar.)

Grid Storage: Is Electricity Stored on the Grid?

Many people are confused when renewable energy experts talk about "storing" electricity on the grid. There's good reason for this.

Technically speaking, renewable energy isn't stored on the electric grid, as it is in batteries. In batteries, electricity is converted to chemical energy for storage. When the electricity is needed, the chemical energy is converted back into electrical energy.

In a grid-connected system, surplus electricity flows onto the grid. It is not physically or chemically stored on the grid, however. Rather, it is immediately consumed by one's nearest neighbors.

That said, electricity fed onto the grid is *effectively* stored thanks to net metering. Here's what I mean: when surplus electricity is backfed onto the grid, a home or business owner is credited for the surplus.

The utility "banks" the electricity. Although it sells the electricity to another customer, the utility keeps track of what has been delivered to the grid, so it can reconcile with the supplier at a later date.

By keeping track of the surplus electricity produced by a net-metered renewable energy system, the utility says, "You've supplied us with *x* kilowatt-hours of electricity. When you need electricity, for example, when the wind is not blowing, our company will supply you with an equal amount at no cost." In a sense, the utility *has* stored the electricity for its customer. Bear in mind, however, when the utility gives back the electricity you've "stored" on the grid, they supply you with electricity most likely generated by coal, nuclear energy, or natural gas. The net effect on your utility bill is the same, however.

Another advantage of grid-connected systems with net metering is that utility customers suffer no losses of electricity when they store surplus electricity on the grid. With battery storage, they do. As described in detail in Chapter 8, when electricity is stored in a battery, it is converted to chemical energy. When electricity is needed, the chemical energy is converted back to electrical energy. As much as 20 to 30 percent of the electrical energy fed into a battery bank is lost due to conversion inefficiencies.

In sharp contrast, electricity stored on the grid comes back in full. If you deliver 100 kilowatt-hours of electricity, you can draw off 100 kilowatt-hours. (I'd be remiss if I didn't point out the grid has losses too, but net-metered customers get 100 percent return on their stored electricity.)

Another notable advantage of batteryless grid-tied systems is that they are greener than battery-based systems. Although utilities aren't the greenest entities in the world, they are arguably greener than battery-based systems. That's because batteries require an enormous amount of energy to produce. Energy is required to mine and refine lead, for example. Batteries must be assembled and shipped long distances, requiring additional energy. Energy production results in pollution, habitat destruction, and a number of other problems. Batteries also contain highly toxic sulfuric acid. Although old lead-acid batteries are recycled, they're often recycled under less-than-ideal conditions. In less developed countries, children are often employed to remove the lead plates along the banks of rivers. Acids from batteries often contaminate surface waters.

Yet another advantage of batteryless grid-connected systems, when net metered, is that they can provide substantial economic benefits. In windy sites, these systems may produce surpluses month after month. If the local utility net meters and pays for surpluses at the end of the month or end of the year at retail, these surpluses can generate income that helps reduce the cost of a wind system and therefore the annual cost of producing electricity.

On the downside, batteryless grid-connected systems are vulnerable to grid failure. That is, when the grid goes down, so does a batteryless grid-connected wind system. Homes and businesses cannot use the output of a batteryless wind or photovoltaic (PV) system when the grid is not operational. Even when winds are blowing, batteryless grid-tied wind energy systems shut down if the grid experiences a problem—for instance, a line breaks in an ice storm or lightning strikes a nearby transformer, resulting in a power outage. Even though the winds are blowing, you'll get no power from your system.

If power outages are a recurring problem in your area, and you want to avoid service disruptions but like the idea of being connected to the grid, you may want to consider installing a standby generator that switches on automatically when grid power goes down. Bear in mind, however, that a standby or backup generator takes many seconds to start and come on line, so your power is interrupted during this time. If you want to avoid this temporary interruption, you could install an uninterruptible power supply (UPS) on critical equipment such as computers. An uninterruptible power supply contains a battery and an inverter. If the utility power goes out, the UPS will supply uninterrupted power until its battery get low. Or, you may want to consider installing a grid-connected system with battery backup. In the latter, you'll need to install a different type of inverter—a battery-based grid-tied inverter—and, of course, a small bank of batteries that provide backup power when the grid goes down.

Grid-connected Systems with Battery Backup

Grid-connected systems with battery backup are also known as battery-based utility-tied systems. These systems ensure a continuous supply of electricity, even when freezing rain wipes out the electric lines that supply you and another 250,000 utility customers in your part of the country.

Grid-connected systems with battery backup contain many of the components found in grid-connected systems, including a turbine, tower, inverter, AC and DC disconnects, main service panel, and utility meter. However, they also contain a few extra or modified components to allow them to operate in conjunction with the grid and batteries. Let me highlight a few differences.

First and foremost, these systems require a *battery-charging* wind turbine. Second, grid-tied systems with battery backup also require an inverter designed to operate off batteries *and* the electrical grid. Third, grid-tied systems with battery backup require a special type of controller, known as a *charge controller*. It controls battery charging, ensuring they are kept in a state of full charge at all times. Fourth, these systems require battery banks, though usually small ones, about one-third of the size of battery banks used in off-grid systems. Fifth, these systems require a different type of inverter, one that is capable of operating on the grid as well as off batteries. They are referred to as *multimode inverters*. The final difference, and this is a big one, is a critical loads panel. It houses the circuits that will be energized when the grid goes down and the system converts to battery power. You may want to take a moment to study Figure 3.6, which shows all of these components.

Although I will discuss all of these components in later chapters, there are a few things you should know upfront. First, because many readers may be thinking about installing a grid-tied system first and then adding batteries for backup at a later date or even going off-grid entirely, it is important to note that you can't use a grid-tied inverter in a battery-based system. You'll need to switch out your inverter. You'll also have to purchase a different type of wind turbine, specifically a battery-charging turbine. (More on this in Chapter 5.)

Batteries for these systems are typically either flooded lead-acid batteries or, more commonly low-maintenance sealed lead-acid batteries. Battery banks are usually sized to provide sufficient storage to run a few critical loads for three to five days until the utility company restores electrical service. Critical loads include your

FIGURE 3.6. Grid-connected Wind System with Battery Backup. Credit: Forrest Chiras.

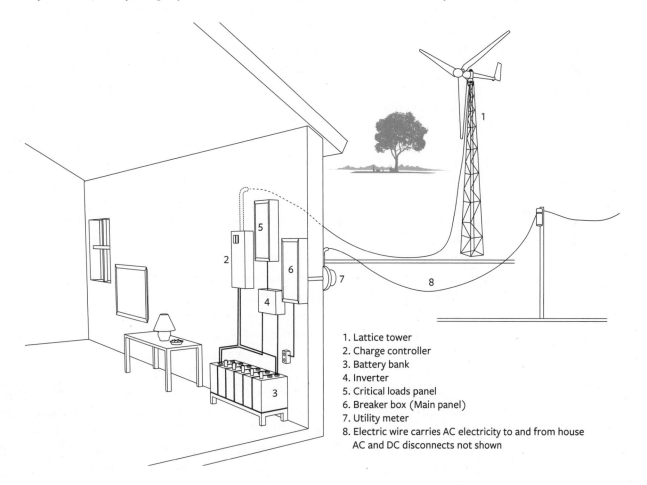

1. Lattice tower
2. Charge controller
3. Battery bank
4. Inverter
5. Critical loads panel
6. Breaker box (Main panel)
7. Utility meter
8. Electric wire carries AC electricity to and from house
   AC and DC disconnects not shown

refrigerator, freezer, a few lights, sump pump, big screen TV (just kidding), and the furnace blower or the pump in a radiant floor or baseboard hot water heating system. You may also need to install a backup generator, just in case. They're typically gasoline-powered.

Another thing to keep in mind: Batteries in these systems serve one function: they provide power in case of outages. They are not used at any other time. You won't be drawing electricity off your batteries at night. If you did, and the grid went down the next day, you might not have adequate backup power left in your battery bank.

Because a fully charged battery bank is essential in the case of a power outage, the battery banks in grid-connected systems are maintained at full charge day in and day out. To reiterate: in battery-based grid-tied systems, batteries are called into duty only when the grid goes down. They're a backup source of power. They're not there to supply additional power, for example, to run loads that exceed the wind system's output. When demand exceeds supply, the grid makes up the difference, not the batteries. When the winds are dead, the grid, not the battery bank, becomes a home or business's power source.

Yet another point worth noting is that maintaining a fully charged battery bank requires a fair amount of electricity. That is to say, a portion of the surplus electricity a wind system generates will be devoted to keeping batteries full.

Batteries require this continual input of electricity because they self-discharge— they lose electricity when sitting idle. You've seen it happen to flashlight batteries or car batteries sitting idle for months. Batteries therefore become a regular load on renewable energy systems.

How much electricity is required to top off a battery bank?

In systems equipped with modern inverters, topping off batteries will consume five to ten percent of a system's output.

Battery banks in grid-connected systems don't require careful monitoring like those in off-grid systems, but it is a very good idea to keep a close eye on them. When an ice storm ravages utility lines in your area and knocks out the power, the last thing you want to discover is that your battery bank is dead.

For this reason, grid-connected systems with battery backup typically include a meter to monitor the total amount of electricity stored in the battery bank. These meters give readings in amp-hours or kilowatt-hours. (See sidebar for definition of Amphours.) You'll learn more about the meters used to monitor batteries in Chapter 8.

Grid-tied systems with battery backup also come with a meter that displays battery voltage. Battery voltage can provide a very general approximation of the amount of energy in a battery—the higher the voltage, the more electricity they contain.

Wind energy systems with battery backup also need a charge controller, shown in Figure 3.6. As you may recall from our discussion of batteryless grid-connected systems, the controller in these systems contains a rectifier. It converts variable AC to DC. DC is then fed into an inverter.

The charge controllers in grid-connected systems with battery backup also contain rectifiers. Besides converting AC to DC, charge controllers, discussed in more detail in Chapter 8, perform other vital functions. For instance, they monitor battery voltage at all times. They use this information to protect batteries from being overcharged—having too much electricity driven into them.

When the charge controller sees that the batteries are fully charged—that is, they have reached the "full" voltage set point and hence are in danger of overcharging—it terminates the flow of electricity to them. Surplus electricity is then fed onto the grid, or, if the grid is not operational, to a diversion or dump load—typically a resistance-type device that accepts the surplus electricity and turns it into heat. Resistance heaters are installed in existing water heaters or as separate space heaters in the basement or a nearby utility room and serve as dump load, making some use of the surplus energy (Figure 3.7).

Charge control is essential to battery-based systems because overcharging batteries can permanently damage their lead plates, dramatically reducing battery life. Most wind turbines designed for battery charging come with an appropriate charge controller.

Batteries also require protection from discharging too deeply, referred to by the techies as overdischarging. Like overcharging, overdischarging damages the lead plates of batteries, dramatically reducing their

Amps and Amp-Hours

Electricity is the flow of electrons through a wire. Like water flowing through a hose, electricity flows through conductors at varying rates. The rate of flow depends on the voltage (defined earlier).

The flow of electrons through a conductor is measured in amperes or amps for short. (An amp is 6.23×10^{18} electrons passing by a point on a conductor per second.) The greater the amperage, the faster the electrons are flowing.

One amp of electricity flowing through a wire for an hour is one amp-hour. This term is also frequently used to define a battery's storage capacity. A flooded lead-acid battery, for example, might store 360 amp-hours of electricity.

FIGURE 3.7. Dump Load. Resistive heaters like this one are used as dumps for surplus electricity from off-grid wind energy systems. Credit: Homepower.com.

productive life. To prevent overdischarging, charge controllers contain a low-voltage disconnect (LVD), although they typically only cover DC loads—that is, circuits that draw DC electricity directly from the battery bank to supply DC appliances and lights. To protect against overdischarging by AC loads, battery-based inverters incorporate a low-voltage AC disconnect. It shuts down the inverter AC output if the battery voltage drops too low.

Low-voltage disconnects in the charge controller and inverter terminate the flow of electricity from the battery bank when the amount of electricity stored in batteries falls to 20 percent of the battery's storage capacity.

In battery-based grid-tied systems, deep discharge is a rare event. Overdischarging typically only occurs during extended utility power outages.

As discussed in depth in Chapter 8, charge controllers also control the charging of batteries. This occurs after battery discharge, for example, during or after periods of power outage when batteries have been extensively used.

Pros and Cons of Grid-connected Systems with Battery Backup

Grid-connected systems with battery backup protect us against utility failures. They enable homeowners to continue to run critical loads—that is, select loads in their homes during power outages—while neighbors grope around in the dark and their food thaws in their freezers and begins to rot in their refrigerators. These systems allow businesses to protect computers and other vital electronic equipment so they can continue operations while their competitors twiddle their thumbs and complain about financial losses.

Although battery backup may seem like a desirable feature, it does have some drawbacks. For one, grid-connected systems with battery backup cost more to install and operate. Flooded lead-acid batteries and sealed batteries used in these and other renewable energy systems are expensive. Remember, too, that a battery bank needs a home. If you are building a new home or office, you also need to add a well-ventilated battery box or battery room that stays warm in the winter and cool in the summer to house your batteries. In addition, unless you install a very large battery bank, your home or business won't be protected against blackouts lasting more than three or four days—unless winds are sufficient to keep your batteries charged. Nor will you be able to meet all of your regular day-to-day power requirements.

Flooded lead-acid batteries also require periodic maintenance and replacement. As explained more fully in Chapter 8, to maintain batteries for long life, you'll need to monitor their fluid levels regularly and fill them with distilled water every few months. If you don't, you'll be sorry. For about ten years, Mick Sagrillo had a backup battery bank in the cellar of his wind-powered home in Wisconsin. This large and costly battery bank was called into service only a couple of times because power outages in his area were infrequent and short-lived, rarely lasting longer than 10 to 20 minutes. Because of this, his batteries were ignored. That is, Mick didn't perform the required maintenance, so when he needed the system, the batteries had lost much of their storage capacity and could provide only a fraction of his family's electricity. He admits that this was a case of "out of sight, out of mind." Because of his experience, he rarely recommends battery backup to clients.

Battery banks also need to be vented to the outside to prevent potentially dangerous hydrogen gas buildup, which can lead to explosions and fires if ignited by a spark. In addition, battery banks need to be replaced periodically, whether you use them or not. Typical batteries used in this system require replacement every five to ten years, at a cost of several thousand dollars each time.

Yet another problem is that battery-based grid-tied systems require a portion of the daily renewable energy production just to keep the batteries fully charged.

Because grid-connected systems with battery backup are expensive and infrequently required, few people install them. My solar-electric system with battery backup, which I installed in 2010, has switched to battery power only twice, following short-term power outages.

When contemplating a battery-based grid-tied system, ask yourself three questions: (1) How frequently does the grid fail in your utility's service area? (2) What are your critical loads and how important is it to keep them running? (3) How do you react when the grid fails?

If the local grid is extremely reliable, you don't have medical support equipment to run or need computers for critical financial transactions, and don't mind using candles on the rare occasions when the grid goes down, why buy, maintain, and replace costly batteries?

In some cases, people are willing to pay for the reliability that a battery bank brings to a grid-connected system. One of my customers in British Columbia buys and sells stocks, bonds, and currency for hedge fund clients. He can't experience down time during active trading, not for a second. As a result, he opted for a grid-connected system with battery backup for his home and office.

See Table 3.3 for a quick summary of the pros and cons of battery-based grid-connected systems.

Table 3.3. Pros and Cons of a Battery-based Grid-tie System

Pros	Cons
Provides a reliable source of electricity	More costly than batteryless grid-connected systems
	Less efficient than batteryless grid-connected systems
	Less environmentally friendly than batteryless systems
	Requires more maintenance than batteryless grid-connected systems

Off-grid (Stand-alone) Systems

Those who want to or must supply all of their needs through wind energy or a combination of wind and some other renewable source without grid backup need to install an off-grid or stand-alone system. As shown in the schematic drawing of an off-grid system in Figure 3.8, this system bears a remarkable resemblance to a grid-connected system with battery backup. There are some noteworthy differences, however.

The first and most obvious is lack of grid connection. That is, there are no power lines running from the house or business to the grid. These systems stand alone. They are self-sufficient power systems in which the wind turbine produces all of the electricity required to meet the needs of a family or business. Off-grid systems rely on batteries to store surplus electrical energy generated during windy periods for use during low- or no-wind periods.

A second major difference between an off-grid system and a grid-connected system with battery backup is that the off-grid system often requires assistance, in the form of a PV array, a micro hydro turbine, or a gasoline or diesel generator, often referred to as a *gen-set*. One or more of these energy sources help make up for shortfalls.

Although backup generators are commonly used in off-grid renewable energy systems, Mick Sagrillo contends that properly sized wind/PV hybrid systems rarely, if ever, require them. In fact, he's retrofitted numerous PV systems with wind generators to avoid the need for generator backup. That said, Ian Woofenden points out that the majority of off-grid systems he's encountered include backup generators. (He lives in the Pacific Northwest where the Sun is often shrouded by clouds.) It takes a very balanced mix of solar and wind resources to avoid the need for a gen-set. "A gen-set also provides redundancy," notes Jim Green. "If a critical component of a hybrid system goes down temporarily, the gen-set can fill in while repairs are made." I agree.

Backup generators are wired directly into inverters. They contain a battery charger that converts the AC electricity from the generator into DC electricity that is used to charge batteries. For this reason, inverters for battery-based systems are called inverter/chargers. (Bear in mind that this is separate from the charge controller, which controls charging of the battery from the wind turbine. The battery charger in the inverter controls charging from an AC source, the generator.)

Off-grid systems require charge controllers to regulate battery charging and protect the batteries from overcharging and overdischarging. As you will see in

Chapter 8, charge controllers in all battery-based systems also divert surplus electricity—when batteries are full—into ancillary dump loads.

Like grid-connected systems with battery backup, an off-grid system requires safety disconnects to permit safe servicing. A DC disconnect is located between the charge controller and the inverter. An AC disconnect is required by Code between the inverter and the main service panel. It should be placed outdoors so that fire departments can cut off AC circuits in case of a house fire.

Off-grid systems also require battery-based inverters that contain a low-voltage disconnect to prevent deep discharge of the battery bank.

As is evident by comparing schematics of the three types of systems, off-grid wind energy systems are the most complex. Moreover, some systems are partially wired for DC—they contain DC circuits that are fed directly from the battery bank to service lights or DC refrigerators. (For a discussion of DC circuits and DC appliances, see the accompanying textbox.)

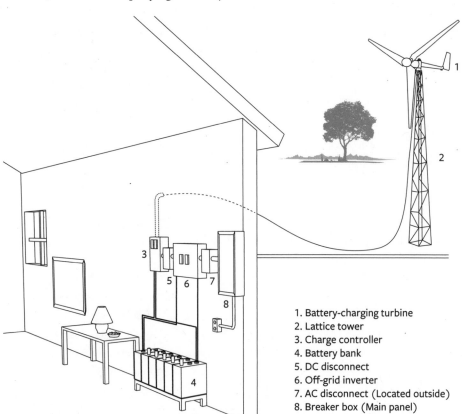

1. Battery-charging turbine
2. Lattice tower
3. Charge controller
4. Battery bank
5. DC disconnect
6. Off-grid inverter
7. AC disconnect (Located outside)
8. Breaker box (Main panel)

FIGURE 3.8. Off-grid Wind System. Most wind turbines in off-grid wind systems produce AC electricity that's converted to DC electricity by the controller. The inverter draws electricity from the batteries, converting DC electricity into AC electricity for household consumption. This drawing shows all of the components of an off-grid wind system. Credit: Forrest Chiras.

DC Circuits in an Otherwise AC World?

Most modern homes and businesses operate on alternating current electricity. However, off-grid homes supplied by wind or solar electricity—or a combination of the two—can be wired to operate partially or entirely on direct current electricity to power DC lights, refrigerators, televisions, and even ceiling fans. Why wire a home or cottage for DC?

One reason is that DC systems do not require inverters. Electricity flows directly out of the battery bank to service loads. This, in turn, can reduce the cost of the system, as household-sized inverters cost $1,000 to $4,000. However, cost-savings created by avoiding an inverter may be offset by higher costs elsewhere. For example, DC appliances typically cost more than AC appliances—considerably more. DC ceiling fans, for instance, cost four times more than comparable AC models. You could pay $300–$350 for a DC model, but $50–$150 for a comparable AC ceiling fan.

DC appliances and electronics are not only more expensive, they are more difficult to find. You won't find them at national or local appliance and electronics retailers.

Many DC appliances are tiny, too. Most DC refrigerators, for example, are miniscule compared to the AC models used in homes. That's because DC appliances are primarily marketed to boat and recreational vehicle enthusiasts, and there's not a lot of room in a boat or recreational vehicle for large appliances.

For these and other reasons, DC-only systems are rare. They are typically installed in remote cabins and cottages that are only occasionally used.

Another reason for avoiding the use of an inverter is efficiency. As you will see in Chapter 7, most inverters for off-grid systems are about 92% to 96% efficient. That is, they consume some energy—4% to 8%—when converting DC to AC and boosting the voltage. Bypassing the inverter with a DC circuit to the water pump or refrigerator reduces this loss. Over the long haul, bypassing the inverter can result in large savings.

Although DC circuits may seem like a good idea, *Home Power*'s Ian Woofenden argues in favor of caution when considering this approach. He notes that water pumping typically does not require a lot of electricity, unless, of course you are irrigating a lawn or large garden or watering livestock. If you are living off-grid, that's not very likely. Ceiling fans also are not that significant in the big picture. However, as Ian says, it's easier to make a case for a DC refrigerator or DC lighting, both more substantial electrical loads in our homes.

Even though modern refrigerators are much more energy efficient than their predecessors, they still are major energy consumers in homes. In off-grid homes, refrigeration often accounts for a substantial amount of a family's daily electrical demand. Because refrigerators consume so much electricity, a DC unit like those offered by Sun Frost or Sun Danzer, can save a substantial amount of energy over the long run (Figure 3.9). Those thinking about an off-grid system, like one of my customers who runs a rustic "hotel" in Tulum, Mexico, with no grid power nearby, may want to consider DC refrigerators and DC freezers like the Sun Danzer chest freezer.

While efficient, DC refrigerators and freezers do not come with the features that many Americans expect, such as automatic defrost or ice makers or cold-water dispensers on the door. They also cost much more than standard or even high-efficiency AC units. For these reasons, I rarely recommend the inclusion of DC circuits in off-grid installations. They're only for a certain type of renewable energy user—people who are trying to wring every possible kilowatt-hour from a small system.

There are other reasons to think twice about DC circuits in an otherwise AC home. For one, low-voltage DC circuits must be wired with larger gauge (thicker) wire. Copper wire is expensive, and the larger the gauge, the more you will spend. Richard Perez, founder of *Home Power* magazine, notes that electrical connections should be soldered, which adds to the cost of installation. DC circuits also require special plugs and sockets that are not readily available and are considerably more expensive than their AC counterparts.

Perez also points out that DC appliances are not typically as reliable or well-built as their AC counterparts. They are, he says, primarily designed for intermittent use in recreational vehicles. DC appliances also wear out more quickly and thus require more frequent replacement.

As a case in point, a DC blender costs about twice as much as an AC model. The blender, Perez jokes, has only two speeds—on and off. Moreover, it can only be ordered from specialty houses by mail or via the internet. The DC blender he bought died after fewer than eight months of use. And if that's not enough to dissuade you from DC appliances, Perez points out that many appliances have no DC counterparts.

Arguing in favor of AC installation, Perez notes in an article in *Home Power* magazine, "The main advantage of using AC appliances is standardization. The wiring is standard—inexpensive, conventional house wiring. The appliances are standard, and are available with a wide variety of features." Furthermore, "the appliances are designed for regular use, and most are reliable and well built."

DC wiring may also need to be installed in metal conduit when it is run through walls, floors, or attics. This adds even more to the cost of a system.

Why metal conduit? DC electricity arcs—that is, it can jump across a gap fairly easily. If a wire is severed, for example, by a nail driven into a wall, sparks "jumping" across the gap could start a fire. (Be sure to check with your local building department or a licensed electrician to determine conduit requirements for DC circuits.)

While the main disadvantage of using AC appliances in off-grid systems is the cost of the inverter and the energy lost due to inversion inefficiency, modern inverters are remarkably efficient. Their operating efficiency is similar to or less than the amount of energy lost in low-voltage DC wiring, especially if the wire runs are long or are not terminated properly. In the final analysis, then, DC may not save any energy at all by bypassing the inverter.

If you are thinking about installing an off-grid system, your best bet is an AC system. Even so, you may want to consider installing a few DC circuits. In my off-grid home in Colorado, which is powered by solar electricity, I used AC electricity almost entirely. However, I wired one DC circuit to power a DC pump that pumps water from my cistern to the house. I added a second DC circuit to power three DC ceiling fans (although, after I priced them I opted for AC fans instead). At the time, I thought that I ought to install a DC refrigerator, but if you shop carefully you can find AC refrigerators that are nearly as efficient as the off-grid DC refrigerators.

You might want to consider installing a DC circuit in the utility room—or wherever the inverter is located—just to power a DC compact fluorescent lightbulb in case of emergency! That way, if the system goes down at night, and you need to find out why, you'll have some light. Be sure to have a spare DC lightbulb on hand, too.

FIGURE 3.9. Sun Frost Refrigerator. In off-grid homes, homeowners typically install the most energy-efficient appliances they can purchase, like this Sun Frost fridge—or they do without. Credit: SunFrost.

Pros and Cons of Off-grid Systems

Off-grid systems offer many benefits, including total emancipation from the electric utility—and Society. They provide a high degree of energy independence that many people long for and some people fear will be necessary should any one of a dozen or so calamities strike. Bear in mind, though, that while you may be independent from the utility, you may not be not totally independent from the world. That is, you'll very likely need to buy a generator and fuel—from large corporations—to maintain your batteries and provide backup power for long, windless periods. Regrettably, generators produce a lot of pollution and cost a lot of money. They are notoriously inefficient. And, they cost money to maintain. To avoid this hassle, you may want to consider installing a solar-electric system for backup.

If designed and operated correctly, and combined with energy-efficiency measures, off-grid wind systems—with proper backup—could provide energy day in and day out for many years.

Off-grid systems do have some downsides. As you might suspect, they are the most expensive of the three options. Large battery banks, supplemental PV systems, and backup generators add substantially to the cost. They also require more wiring. You will also need space to house battery banks and generators. (I had to build an insulated, soundproof ventilated shed to house my backup generator to appease neighbors who complained about the noise.) As noted earlier, batteries also require periodic maintenance and replacement every seven to ten years, depending on the quality of batteries you buy and how well you maintain and treat them.

Table 3.4. Pros and Cons of Off-grid Systems

Pros	Cons
Provide a reliable source of electricity	Generally the most costly wind energy system
Provide freedom from the utility grid	Less efficient than batteryless grid-connected systems
Can be cheaper to install than grid-connected systems if located more than 0.2 miles from grid	Require more maintenance than batteryless grid-connected systems (you take on all of the utility's operation and maintenance jobs and costs)

Although cost is a major downside, there are times when off-grid systems cost the same or less than grid-connected systems—for example, if a home is located more than a few tenths of a mile from the electric grid. Under such circumstances, it can cost more to run electric lines to a home than to install a small off-grid system. (I'll talk more about this in later chapters.)

When installing an off-grid system, remember that *you* become the local power company, and your independence comes at a personal cost. It also comes at a cost to the environment, because the production of lead-acid batteries is not benign. As noted earlier, although virtually all lead-acid batteries are recycled, battery production is responsible for considerable environmental degradation. Mining and refining the lead, for instance, are fairly damaging. Lead production and battery recycling are often carried out in poor countries with lax or nonexistent environmental policies. According to Mick, they are responsible for some of the most egregious pollution and health problems facing poorer nations. So, think carefully before you decide to install an off-grid system. For more on this topic, see the accompanying sidebar.

To Stand Alone or Not to Stand Alone?

Many people speak to me about going off grid by installing a PV system, even though their homes or businesses are currently grid-tied. While this may sound like a glorious way to live your life, the off-grid option comes at a price. When Mick Sagrillo consults with individuals who want to go off grid for philosophical reasons, he cautions them to consider the ramifications of this decision, especially if they are philosophically opposed to utilities and concerned about the environmental impacts of grid-generated electricity. He does not at all condone how utilities operate, but from a purely environmental perspective, he thinks it is extremely difficult to justify a battery bank and gen-set system over grid connection.

I discussed the downsides of batteries earlier, but haven't mentioned the generator. The electric grid delivers electricity generated by coal-fired power plants at about 30%–33% efficiency and nuclear power plants at 20% efficiency. A Honda gen-set charging a battery bank operates at about 5% efficiency, Sagrillo contends. The emissions from that coal-fired plant are regulated by the EPA—more or less. Although there are also emission regulations for small engines (manufactured since September 1, 1997), the emissions per kilowatt-hour of electricity are far greater from the backyard generator than from the coal-fired power plant. In fact, if everyone had a gen-set in their backyard to meet their electrical demands, we'd all suffocate. Visit Cali, or Beijing, or any number of developing countries where generators are used to produce electricity and you'll see, smell, and choke on the result.

If you are thinking about going off grid, the responsible approach might be to install a hybrid system in which solar electricity is supplemented by some other clean, renewable energy technology such as wind or micro hydro.

Hybrid Systems

As pointed out in this chapter, individuals and businesses have three basic options when it comes to wind energy systems. Each of these systems can be hybridized—that is, they can be designed to include another renewable energy system, such as a solar-electric system (Figure 3.10).

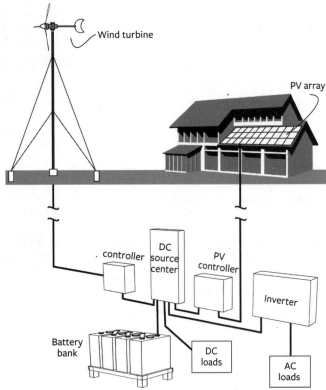

Hybrid renewable energy systems are extremely popular among homeowners in rural areas. In fact, most residential off-grid wind systems in use in North America are hybrids that combine solar electricity with wind.

Wind and PV are a marriage made in heaven in many parts of the world. In many locations in North America, like the Great Plains, winds are consistent and strong throughout the year. Consider Greensburg, Kansas, the small town destroyed by a tornado in the spring of 2007. Table 3.5 presents average monthly wind speeds in Greensburg at 120 feet above the flat short-grass prairie. As you can see, Greensburg experiences robust winds year-round.

In most other locations, winds vary throughout the year. They tend to be strongest in the fall, winter, and early spring—from October or November through March or April. Table 3.6, shows the average wind speed at a farm in central Missouri. As you can see, the strongest winds occur from October through May. During these months, a properly sized wind generator can meet most of a family or business's needs. June through September is less windy.

FIGURE 3.10. Hybrid System. To meet their needs, some homeowners and business owners install hybrid systems consisting of two or more energy sources, often wind and solar electricity. Bear in mind that adding a second system can be quite costly. It may make more sense to install a larger wind system rather than install a hybrid wind system and solar-electric system. Credit: Anil Rao.

Table 3.5. Ten-Year Monthly Average Wind Speed in Greensburg, Kansas, at 120 Feet

	Jan	Feb	Mar	Apr	May	Jun	Jul	Aug	Sep	Oct	Nov	Dec	Annual Average
meters/second	6.55	6.63	7.50	7.60	6.97	6.60	6.26	6.01	6.30	6.36	6.43	6.44	6.64
miles/hour	14.65	14.83	16.80	17.00	15.60	14.76	14.00	13.44	14.09	14.23	14.38	14.41	14.85

Table 3.6. Ten-Year Monthly Average Wind Speed Near Gerald, Missouri, at 120 feet (Lat 38.325, Long 91.297)

	Jan	Feb	Mar	Apr	May	Jun	Jul	Aug	Sep	Oct	Nov	Dec	Annual Average
meters/second	5.96	5.93	6.50	6.31	5.27	4.83	4.40	4.31	4.64	5.16	5.67	5.82	5.40
miles/hour	13.33	13.26	14.54	14.12	11.79	10.8	9.84	9.64	10.38	11.54	12.68	13.02	12.08

Table 3.7. Solar Resource near Gerald, Missouri, measured in kWh/m² per day (Lat 38.325, Long 91.297)*

	Jan	Feb	Mar	Apr	May	Jun	Jul	Aug	Sep	Oct	Nov	Dec	Annual Average
Tilt 38	3.66	3.86	4.71	5.43	5.28	5.50	5.75	5.66	5.48	4.79	3.27	2.95	4.70

* This data represents solar energy striking a collector mounted at an optimal angle for this location.

Winds continue to blow, but less frequently and less forcefully. Fortunately, sunshine is more abundant during this period, as shown in Table 3.7. Installing a solar-electric system would supplement the wind-generated electricity. It could provide the bulk of the electricity during this period while the wind turbine played a subsidiary role, backing up the PV system. Together, they can supply 100 percent of the annual electrical demand. This complementary relationship is shown graphically in Figure 3.11.

Be sure to run the math very carefully on a hybrid system. Installing two systems—both a wind and a solar-electric system—may be more expensive than simply installing one larger system—for example, a larger wind turbine or a large solar-electric system.

Because solar and wind resources are often complementary, hybrid systems provide a more consistent year-round output than either wind-only or PV-only systems.

Sized correctly in areas with a sufficient solar and wind resource, a hybrid wind/PV system can not only provide 100 percent of one's electricity, it may eliminate the need for a backup generator. What is more, hybrid systems often require smaller solar-electric arrays and smaller wind generators than if either were the sole source of electricity.

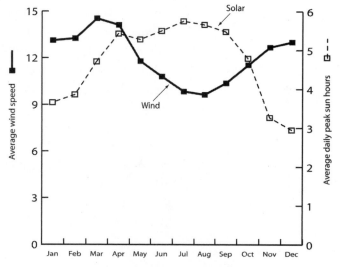

FIGURE 3.11. The Complementary Nature of Wind and Solar. Wind and solar energy often complement each other and can be combined to create a reliable, year-round source of electricity. Note how well solar makes up for the reduction in wind during the summer months at our site.

If the combined solar and wind resource is not sufficient throughout the year or the system is undersized, a hybrid system may require a backup generator. A gen-set will supply electricity during periods of low wind and low sunshine. It is also used to maintain batteries in peak condition, as discussed in Chapter 8, and it may allow installation of a smaller battery bank.

Choosing a Wind Energy System

Although you have a lot of choices when it comes to installing a wind system, by far the cheapest and simplest option is a batteryless grid-connected system.

If you are installing a wind system on an existing home that is already connected to the grid, it's generally best to stay connected. Use the grid as your battery bank. Although you may encounter an occasional power outage that shuts your system down, in most places these are rare and transient events.

If you need backup power, by all means consider a grid-tied system with battery backup. They cost more than grid-tied systems, but provide peace of mind and security.

Although more expensive than all other systems, off-grid systems are often the system of choice for people in remote rural locations. If you are building a new home and you are more than a few tenths of a mile from existing power lines, connecting to the grid could be expensive. Although some utility companies foot the bill for line extension, some will charge you to run the utility line to your home. You don't have to live that far from the grid for an off-grid system to make sense. In some locations, a quarter-mile grid connection is costly enough to justify an off-grid system.

Be sure to check with your utility. "Utility policies vary considerably when it comes to line extension costs," notes Jim Green. Hook-up fees can be upward of $50,000 if you live half a mile from the closest electric lines. I've known people who have paid $65,000 to run a line a mile to two small cabins. They could have installed an off-grid wind or solar system to meet the meager needs of these cabins much less expensively.

CHAPTER 4

Wind Site Assessment

Wind is an abundant and clean energy resource available in many parts of the world for commercial and residential use. Because it is renewable, this vast resource will be available to humankind for as long as its source—the Sun—continues to warm our planet. Unfortunately, wind energy systems are not suitable for all locations. If you live in a city, suburb, or small town, for instance, a small wind system is probably not going to work, for reasons you will soon learn. If you live in a heavily forested area, wind can be a tricky proposition. So before you invest your hard-earned money in a wind system, it is important to determine whether it makes sense for you to invest in one.

For many people, making sense means making economic sense. Others are content to invest solely for environmental reasons. For others, a wind energy system might be a fun and interesting hobby, one that pays a little back. I have had clients in less-than-optimal sites for whom my wind site and economic assessments indicated that a wind system would provide a return on investment well under one percent per year, but who decided to install the system anyway. They wanted a wind turbine and were willing to put their money where their values were. They made a commitment to the future of their children and grandchildren and the rest of humanity.

As a rule, the decision to install a wind energy system is based on a combination of factors. I talked about some of them in Chapter 1 when I discussed the benefits of wind turbines.

If the economics of a wind system is your primary or perhaps only focus, the decision to invest time and money on a wind system requires an understanding of three key factors: (1) your electrical energy requirements—how many kilowatt-hours of electricity you need; (2) the number of kilowatt-hours of electricity a wind

The Economics of Wind

Several factors help to make small wind economically viable and can make it a more profitable venture. They include: (1) high electricity rates, (2) financial incentives like rebates or tax credits from utilities or governments, (3) a windy site with few obstructions, (4) a site that permits placement of a turbine on a tall tower, (5) favorable net metering policies (notably, annual net meter with net excess generation reconciled at retail rates), and (6) a long-term perspective.

In California, a state that is trying to generate half of its electricity from renewable energy resources by 2030, generous financial incentives are available. Combined with high electrical rates—ranging from 20 cents to 40 cents per kilowatt-hour—small wind system often make economic sense. In Hawaii, where electricity costs 25 cents per kilowatt-hour, wind makes sense in any good wind site even *without* incentives. In Germany, where electricity costs 75 cents per kilowatt-hour, wind is an even better investment. On the other hand, in some parts of Washington state, where hydroelectricity costs a paltry three to four cents per kilowatt-hour, it's nearly impossible for wind to compete economically. The same can be said for British Columbia, where electricity costs just over six cents per kilowatt-hour.

turbine could produce at your site; and (3) the cost of installing and maintaining a system to meet your needs. With this information, you can compare wind-generated electricity to utility power or some other renewable energy source, such as solar electricity or micro hydro. Even if money isn't a primary driving force, most of us want to know if a wind energy system—or a hybrid wind energy system—represents an intelligent investment. This chapter will help you analyze the investment decision.

To begin, let's explore how you determine electrical demand. This will help you determine how much electricity you'll need to generate from a wind-electric system. Although I'll focus principally on residential demand, this discussion also pertains to a host of other applications, among them cabins, cottages, offices, farms, and ranches.

Assessing Electrical Demand

Determining monthly electrical consumption is fairly straightforward in existing homes and businesses. Determining electrical demand in a new home—either one that's just been built or one that is about to be built—is much more difficult. Let's look at this challenge first.

One method used to estimate electrical consumption in a new home is to base it on the electrical consumption of your existing home or business. For instance, suppose you are building a brand new home that's the same size as your current home. If the new home will have the same amenities and the same number of occupants, electrical consumption could be similar to your existing home.

If, however, you are building a more energy-efficient home and are installing much more energy-efficient lighting and appliances and incorporating passive solar heating and cooling (all of which I highly recommend), electrical consumption could easily be 50%, perhaps 75%, lower than in your current home. If that's the case, adjust your projected electrical demand to reflect your new, more efficient lifestyle.

Another way to estimate electrical consumption is to perform a load analysis. A load analysis is an estimate of electric consumption based on the number of electronic devices in a home, their average daily use, and energy consumption. It is a lot more difficult to calculate than you would expect.

To perform a load analysis, a homeowner begins by listing all the appliances, lights, and electronic devices that will be used in his or her new home. Rather than list every lightbulb separately, however, you may want to lump them together by room. I like to work with clients one room at a time, using a spreadsheet I've prepared for each project. Or, you can use a worksheet like the one shown in Table 4.1. Similar worksheets can be found online at Northern Arizona Wind and Sun's website (solar-electric.com) and other websites.

Once you've prepared a complete list of all the devices that consume electricity, your next assignment is to determine how much electricity each one uses. The amount of electricity consumed by an electronic device can be determined by consulting a chart like the one in Table 4.2. Charts such as this list typical wattages for a wide range of devices. Detailed listings are available online. Better yet, log on to WE Energies's website. It contains nifty energy calculator that allows you to estimate electric consumption room by room (find it at webapps.we-energies.com/appliancecalc/appl_calc.cfm).

For more accurate data, I strongly recommend that you check out the nameplates on the appliances and electronic devices you'll have in your new home. The nameplate is a sticker or metal plate that lists the unit's electrical consumption. The measure you are looking for is watts. As you may recall, wattage is a measure of instantaneous power consumption. Manufacturers sometimes list amps and volts. If the nameplate lists amps, but not watts, calculate wattage by simply multiplying the amps and volts (amps × volts = watts).

Multiplying amps by volts yields the wattage of many electronic devices, among them resistive devices, such as electric heaters and electric stoves. (A resistive

Table 4.1. Load Analysis

Individual Loads	Quantity	× Volts	× Amps	=	Watts AC	DC	× Use (hrs/day)	× Use (days/wk)	÷ 7 days	=	Watts Hours AC	DC
									7			
									7			
									7			
									7			
									7			
									7			
									7			
									7			
									7			
									7			

AC Total Connected Watts: _____   DC Average Daily Load: _____

DC Total Connected Watts: _____

AC Average Daily Load: _____

device is one in which electricity flows through a metal heating element.) The metals chosen for these devices resist the flow of electrons. This, in turn, results in the production of light (they often glow) and heat. Multiplying amps by volts also works for universal motors. Universal motors are typically found in smaller devices such as vacuum cleaners, blenders, and small electric tools. For devices with induction motors, such as fans, washing machines, clothes dryers, dishwashers, pumps, and furnace blowers, multiplying amps by volts significantly overestimates the wattage. For these devices, multiplying watts (the product of amps and volts) by 0.6 will yield a better estimate of wattage. (For a description of the rationale behind this derating, see the accompanying sidebar, "Real Power, Apparent Power, and Power Factor.")

For the purposes of calculation, some experts reduce the wattage of other de-vices, for example, televisions, stereos, and power tools, below their listed power consumption as well. They do this because the listed wattage is not representative

Table 4.2. Average Electrical Consumption of Common Appliances

General household		Kitchen appliances		Entertainment	
Air conditioner (1 ton)	1500	Blender	350	CB radio	10
Alarm/Security system	3	Can opener (electric)	100	CD player	35
Blow dryer	1000	Coffee grinder	100	Cellular telephone	24
Ceiling fan	10–50	Coffee pot (electric)	1200	Computer printer	100
Central vacuum	750	Dishwasher	1500	Computer (desktop)	80–150
Clock radio	5	Exhaust fans (3)	144	Computer (laptop)	20–50
Clothes washer	1450	Food dehydrator	600	Electric player piano	30
Dryer (gas)	300	Food processor	400	Radio telephone	10
Electric blanket	200	Microwave (.5 ft³)	750	Satellite system (12 ft dish)	45
Electric clock	4	Microwave (.8 to 1.5 ft³)	1400	Stereo (avg. volume)	15
Furnace fan	500	Mixer	120	TV (31.5-inch color)	22–30
Garage door opener	350	Popcorn popper	250	TV (40-inch color)	28–37
Heater (portable)	1500	Range (large burner)	2100	TV (48-inch color)	35–55
Iron (electric)	1500	Range (small burner)	1250	VCR	40
Radio/phone transmit	40–150	Trash compactor	1500		
Sewing machine	100	Waffle iron	1200	**Tools**	
Table fan	10–25	Lighting		Band saw (14″)	1100
Waterpik	100	Incandescent (100 watt)	100	Chain saw (12″)	1100
		Incandescent light (60 watt)	60	Circular saw (7¼″)	900
Refrigeration		Compact fluorescent	16	Disc sander (9″)	1200
Refrigerator/freezer	540	(60 watt equivalent)		Drill (¼″)	250
22 ft³ (14 hrs/day)		Incandescent (40 watt)	40	Drill (½″)	750
Refrigerator/freezer	475	Compact fluorescent	11	Drill (1″)	1000
16 ft³ (13 hrs/day)		(40 watt equivalent)		Electric mower	1500
Sun Frost refrigerator	112			Hedge trimmer	450
16 ft³ (7 hrs/day)		**Water Pumping**		Weed eater	500
Vestfrost refrigerator/freezer	60	AC Jet pump (¼ hp)	500		
10.5 ft³		165 gal per day, 20 ft. well			
Standard freezer	440	DC pump for house	60		
14 ft³ (15 hrs/day)		pressure system (1–2 hrs/day)			
Sun Frost freezer	112	DC submersible pump	50		
19 ft³ (10 hrs/day)		(6 hours/day)			

of the typical run wattage—the wattage an appliance will draw when in normal operation. That's because the wattage listed on an appliance nameplate is the power the device draws at maximum load. Most devices rarely operate at maximum load, so the nameplate power overstates the actual power used. Reducing the nameplate wattage by 25% is a reasonable adjustment for such devices, although typical operating wattage may be even lower.

Real Power, Apparent Power, and Power Factor

To calculate power consumption (watts) of electrical resistance devices, such as electric stoves or computers, simply multiply the voltage by the amps (power [watts] = volts × amps). However, for devices that have induction motors (listed in the text), transformers, and fluorescent lamp ballasts, the product of volts and amps is actually a bit higher than the actual wattage. In such instances, amps × volts is known as the *apparent power*. The discrepancy is due to the fact that inductors and capacitors store energy as *current*, and voltage waveforms *alternate* from positive to negative. This stored energy causes the current waveform to be out of step with the voltage waveform. When the current waveform is out of step with the voltage waveform, more current is required to deliver the same amount of power. The factor that accounts for this difference between apparent power and real power is the *power factor*.

Power factor is the real power divided by the apparent power. Power factor = watts/(volts × amps). Power factor is a number between zero and one (although it can be expressed as a percent). When calculating the real power (watts) for a device that has inductors or capacitors, you must multiply voltage × amps times power factor. Power = volts × amps × power factor. Unfortunately, you won't find power factor listed on appliance nameplates, but a reasonable estimate for most household devices containing inductors or capacitors is 0.6.

Be sure, when determining the run wattage of laptop computers that plug into a power cube (charger or transformer) to use the wattage listed on the power cube, not the device. The same applies to cordless drills, cordless phones, and electronic keyboards.

A more accurate way to determine the wattage (power) of an electronic device—and one I greatly prefer—is to measure it directly using a meter like those shown in Figure 4.1. To use it, plug the meter into an electrical outlet and then plug the appliance into the outlet on the face of the device. A digital readout indicates the instantaneous power (watts). When measuring a device that cycles on and off, such as a refrigerator, or one that has a varying load, leave the watt-hour meter connected for a week or two. The meter will record the total energy used during this period in watt-hours and will tell you how many hours you have been recording data. You can then calculate the number of kilowatt-hours the device consumes in a 24-hour day.

Power consumption can also be determined by checking spec sheets for various electronic devices. You can even go online to find them. Most will list key electrical parameters.

After you have determined the wattage of each electrical device, you must estimate the number of hours each one is used on an average day and how many days each device is used during a typical week. From this information, you calculate the weekly energy consumption of all devices in your home or business. You then divide this number by seven to determine your average daily consumption in watt-hours. You can multiply this number by 365 to determine annual energy consumption.

Load analysis may seem simple at first glance, but it is fraught with problems. One shortcoming I've encountered is that it is often difficult for clients to estimate how long appliances run on a typical day. How many hours or minutes does your toaster or blender run each day? Is it three minutes, five minutes, or ten? In addition, while some people may be able to provide a fairly accurate estimate of how long the kitchen lights are on, they haven't the foggiest idea how long the refrigerator runs or how many minutes they run the microwave each day. Most people tend to underestimate TV time, too. And what about the kids? Do they leave the lights on when you're not around? How many hours of television do they watch each day? How many hours a day do they spend on their computers?

Plug-in Watt-Hour Meters

The Kill A Watt and Watts Up? meters make it easy to measure power (watts) and energy (watt-hours) used by plug loads: lights, refrigerators, freezers, computers, TVs, etc. Just plug the meter into a regular 120-volt outlet and plug the appliance into the meter. Both meters read power (watts), volts, amps, and elapsed time (used for calculating average power use over time). These meters, however, are not designed for measuring electrical consumption by 240-volt appliances.

Power meters are useful for estimating load so you can size a wind energy system, but they are also useful for gaining an appreciation for the energy demands of household appliances. Knowing which devices consume the most energy in a home may help you devise a strategy to reduce energy consumption. Surveying the energy use of household appliances could be a great educational project for children.

Both meters are available in a range of models, some with advanced features. The Kill A Watt meter's list price is about $40; the Watts Up? meter's list price is about $135. Both are available online at a discount. As an alternative to purchasing a meter, some public libraries and utilities have plug-in watt-hour meters available for loan.

FIGURE 4.1. (*a*) The Kill A Watt and (*b*) Watts Up? meters can be used to measure wattage of household appliances and electronic devices. Of the two, the Watts Up? is the more sensitive and allows for measurement of tiny phantom loads as well. Credit: (*a*) Dan Chiras, (*b*) Electronic Educational Devices.

Another problem with this process is that run times vary by season. Electrical lights, for example, are used much more in the winter, when days are shorter, than in the summer. Furnace blowers operate a lot during the winter, but not at all or very infrequently the rest of the year. So what's the average daily run time for the furnace blower?

Another problem with this approach is that many electronic devices draw power when they're off. Such devices are known as *phantom loads*. They include television sets, VCRs, satellite receivers, cell phone and laptop computer chargers, and a host of other common household devices. Any electronic device that has an LED light that shines all the time or comes with a remote and instant-on feature will constitute a phantom load. At times, phantom loads can be quite significant. In years past, I have found that satellite receivers draw nearly as much power when they're off as when they're on. My new DirecTV satellite receiver uses 24 watts when it is off, and 26 when it is on. Phantom loads typically account for 5% to 10% of the monthly electrical consumption in US homes. If they're not factored into the load analysis, estimates can be off.

Phantom loads may be quite small, but they add up over time. Plug-in watt-hour meters can be used to measure phantom loads—if you let them run long enough for the kWh reading to register on the display.

Because of these and other problems, homeowners often grossly underestimate their electrical consumption. So, if you run a load analysis to estimate power consumption, be sure to include phantom loads and be generous with your run-time estimates.

As a side note, you may have noticed that the worksheet shown in Table 4.1 has a column labeled DC. This is for those folks who want to install an off-grid system and power some or all of their loads with DC electricity. DC wattages are also the product of amps × volts. A device that uses 1 amp at 48 volts requires 48 watts.

Once you've calculated daily electrical energy use, it's time to size a wind system, right?

Actually, no. Before you size a system, it's a very good idea to look for ways to reduce electrical use through energy-efficiency and conservation measures. That's because the lower the energy demand, the smaller the turbine you'll need. The smaller the turbine, the less you'll spend. Smaller turbines also require fewer resources to produce, so you're helping reduce your environmental impact, including your carbon footprint.

Money spent on energy-efficiency measures may also have a better return on investment than money spent on a wind energy system. Because efficiency is both economically and environmentally superior to increasing the capacity of a renewable energy system, some really conscientious installers recommend energy conservation and efficiency measures first. If you agree to pursue these measures, you may be able to downsize your system, which could save you thousands of dollars. How do you determine the most cost-effective measures to reduce electrical consumption?

One way is to hire a professional solar site assessor. For a relatively small fee, usually $300 to $600, depending on travel time and complexity of the analysis, an energy auditor will provide a list of ways you can trim your energy demand. Professional small wind system assessors can do the same, if you can find one in your area. They will also objectively evaluate the potential of your site for a wind energy system, tell you what kind of system will best meet your needs (grid-tied or off-grid, for instance), recommend height and placement of the tower, the turbine you should consider installing to meet your needs, and the location of the rest of the equipment. He or she will also provide an approximate cost and perform an economic analysis. (To locate a solar small wind site assessor in your area, visit the Midwest Renewable Energy Association's website. They used to be the certifying body for small wind site assessors in the United States; although they no longer provide certifications, they still list previously certified assessors like myself. To my knowledge, there is no national certifying organization for wind site assessors.)

If you just want a home energy audit, hire a home energy auditor—they are much easier to find in most urban areas. A qualified home energy auditor will perform a more thorough energy analysis of your home and make recommendations on ways you can reduce your demand. They will also prioritize energy-saving measures. They won't be able to give you any advice on a wind system, however. Alternatively, you can perform your own home energy audit. For guidance, check out my book, *Green Home Improvement*. It contains detailed instructions on home energy audits.

Most homeowners contact wind system installers to evaluate their site and make recommendations. In my experience, very few installers will offer advice on energy conservation or energy efficiency. They are there to sell you a system—and the bigger the better. Some inexperienced or unscrupulous installers will also recommend installing wind systems in less-than-optimum sites—for example, on short towers in sites without sufficient wind to make your investment worthwhile. That's why an independent small wind site assessor can be such a good investment. They are working for you and not lining their pockets with the profit they make on an underperforming wind turbine.

Conservation and Efficiency First!

So, before you invest in a wind energy system, the first step should be to make your home—and your family—as energy efficient as possible. Even if you're an energy miser, you may be able to make significant cost-saving cuts in energy use.

Waste can be slashed many ways. Interestingly, though, the ideas that first come to mind for most homeowners tend to be the most costly: new windows and energy-efficient washing machines, dishwashers, furnaces, or air conditioners (Figure 4.2). While vital to creating a more energy-efficient way of life or business, they're the highest fruit on the energy-efficiency tree—and the most expensive.

Before you spend a ton of money on new appliances or better windows, I strongly recommend that you start with the lowest-hanging fruit. These are the simplest and cheapest improvements and yield the greatest energy savings at the lowest cost.

Huge savings can be achieved by changes in behavior. You've heard the list a million times: turning lights, stereos, computers, and TVs off when not in use. Turning the thermostat down a few degrees in the winter and wearing sweaters and warm socks. Turning the thermostat up in the summer and running ceiling fans. Opening windows to cool a home naturally, especially at night, and then closing the windows in the morning to keep heat out during the day. Drawing the shades or blinds on the east, south, and then west side of your house as the summer Sun moves across the sky can help reduce cooling costs. All these changes cost nothing, except a little of your time, but they can reap enormous savings—not just in your monthly energy bill, but also in the cost of a wind energy system.

Other low-hanging, high-yield fruit includes boosting insulation levels and weather-stripping and caulking to seal leaks in the building envelope—tightening up and bundling up our homes and workplaces. These measures reduce heat loss in the winter and heat gain in the summer. They not only reduce fuel bills, they dramatically increase comfort levels. Unfortunately, they are very rarely viewed as a high priority.

Because our homes are so leaky, sealing up the

FIGURE 4.2. Energy-efficient Washing Machine. Spending a little extra for an energy-efficient front-loading washing machine like this will reduce the size of your solar system. To dry clothes, though, consider using a solar clothes dryer (commonly called a *clothesline*). Credit: Frigidaire.

leaks can make a huge difference in energy costs and comfort. If you add up all the tiny leaks in a building envelope around doors and windows and where plumbing and electrical penetrate a home, they'd be equivalent to a 3-foot by 3-foot (0.9 × 0.9 meter) window open 24 hours a day, 365 days a year. Yes, that's right. No one in their right mind would leave a window open all year long, but that is, in essence, what we're doing by not sealing up all those energy leaks in our homes.

Once you've tackled these simple, relatively inexpensive steps, it's time to consider more costly, slightly bigger-ticket items. The next step is to boost the insulation in your home. Insulate windows with insulated shades. Insulate attics, floors above crawl spaces or garages, and walls. An energy consultant can help you figure out the best ways to seal and insulate your home.

Another big-ticket item that can help is replacing your refrigerator. In many homes, refrigerators are responsible for a staggering 25 percent of the total electrical consumption. If your refrigerator is old and in need of replacement, unplug the energy hog, recycle it, and replace it with a more energy-efficient model.

Thanks to dramatic improvements in design, refrigerators on the market today use significantly less energy than refrigerators manufactured 15–20 years ago. An efficient 21 cubic foot fridge today will consume around 365 kWh per year. A 20-year-old fridge could easily consume four times that amount. Whatever you do, though, don't lug your old fridge out to the garage or take it down to the basement and use it to store an occasional case of soda or beer. It will rob you blind! The same goes for old freezers.

Waste can also be reduced by replacing energy-inefficient electronics with newer, considerably more efficient Energy Star televisions, computers, and stereo equipment (Figure 4.3). You can also trim some of the fat from your energy diet by installing more energy-efficient lighting, such as compact fluorescent lights or longer-lasting and even more efficient LED lights. (For more on Energy Star appliances see the accompanying textbox, "Energy Star—and Beyond.")

Although efficiency has been the mantra of energy advocates for many years, don't discount its importance just because the advice has grown a bit threadbare. As it turns out, very few people have heeded the persistent calls for energy efficiency.

FIGURE 4.3. Energy-efficient Stereo with Energy Star Label. This label assures you that the product you're looking at is one of the most energy efficient in its category.

Energy Star—and Beyond

Although all new appliances and electronics must comply with federal energy-efficiency regulations, those that exceed these standards display an Energy Star label. This label indicates that the appliance or electronic device that displays it is one of the most efficient in its product category. So, when it is time to replace a television or stereo—even a cordless phone or laptop computer—always look for models with the Energy Star label. As Table 4.3 shows, Energy Star refrigerators use 20% less energy than is federally mandated. Dishwashers use 41% less and clothes washers use 37% less energy than federal law requires.

Although buying an Energy Star-rated appliance or electronic device is a good idea, you can do better, sometimes much better. How?

Go to the Energy Star website and click on the appliance or electronic device listing. Here you'll find a list of all Energy Star-qualified models. The list indicates the percentage by which each device exceeds the minimum efficiency for that type of appliance.

As Table 4.3 shows, the best appliances on the list are considerably more energy efficient than other Energy Star-qualified products.

If you shop from the Energy Star list, you will do well, but if you search the list for the most efficient models, you can do much better.

For an up-to-date list of energy-efficient appliances, US readers can log onto the EPA's and DOE's Energy Star website at www.energystar. gov. Click on appliances. Or, *Consumer Reports* has an excellent website that lists energy-efficient appliances as well. Their site also rates appliances on reliability, another key factor. Canadian readers can perform a search for a list of Energy Star-certified products in Canada or log on to nrcan.gc.ca/energy/products/energystar/why -buy/13631.

You can also compare major appliances like refrigerators and freezers by checking out the yellow Energy Guide posted on these devices in stores. The Guide will tell you how much electricity a particular appliance will use and how the model you are looking at compares to other models in that category. Figure 4.4 shows an example of an Energy Guide. In Canada, the same yellow tags are also posted on appliances, but it's called an EnerGuide.

When shopping for TVs and other electronics look for the Energy Star logo, but turn the darn things around and read the label on the back to check out consumption in wattage. That way, you can compare Energy Star-certified products in the store. You would be amazed at the variation. I found a 50-inch (127-cm) Toshiba LED flat screen that uses 61 watts. That's only 12 watts more than my efficient laptop.

Table 4.3. Energy Star Performance (% better than federal mandated efficiency)

Appliance Type	Energy Star Criteria	Best on the List
Refrigerators	20%	53%
Dishwashers	41%	147%
Clothes Washers	37%	121%

And, many who have made changes have not fully tapped the potential savings.

Readers interested in learning more about making their homes energy efficient may want to read the chapters on energy conservation in one of my books: *Green Home Improvement* or *The Homeowner's Guide to Renewable Energy*.

Assessing Your Wind Resource

Once you have executed a strategy to make your home or business—and its occupants—more energy efficient and have recalculated energy consumption, it is time to assess the potential of your wind resource. How much wind is available to you and when it is available? From this information, you can select a wind turbine that will produce enough electricity to meet your needs. After that, you can evaluate system costs and run a financial analysis to determine the cost of electricity generated by your system and a simple return on investment.

An assessment of available wind resources can be made using one or a combination of the following techniques: (1) direct measurement, (2) local airport and weather service data, (3) wind maps, and (4) online sources. You can also use other information to help you confirm data from other sources, which I will discuss shortly. Professional small wind site assessors rely principally on online sources.

Direct Measurement

By far the most accurate way of assessing wind is direct monitoring—measuring the wind speed at the tower height on the site you are considering. "Direct monitoring by a wind resource measurement system provides the clearest picture of the available resource," note the authors of the US Department of Energy's booklet, *Small Wind Electric Systems*. For best results, they contend, wind speed should be monitored over a period of at least one year. Given the variability of wind from year to year, however, the most useful results are obtained by two or more years of monitoring.

The best way to measure wind resources on site is by using a pole-mounted anemometer located on the site at the height of the future wind turbine. Anemometers come in two basic varieties. A cup anemometer consists of small cups mounted

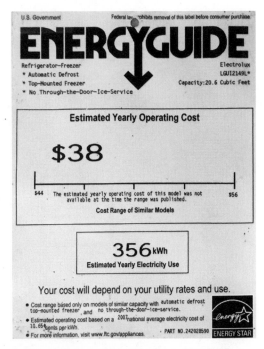

FIGURE 4.4. Energy Guide Tag. When shopping for appliances such as refrigerators and freezers, be sure to look for the Energy Guide, like the one shown here. These guides indicate how much electricity an appliance will use in one year and how the appliance you are looking at compares to others in its product category. This one is for the refrigerator/freezer that we purchased for our home; it uses the least energy of all the Energy Star-rated refrigerators in its category. Credit: Dan Chiras.

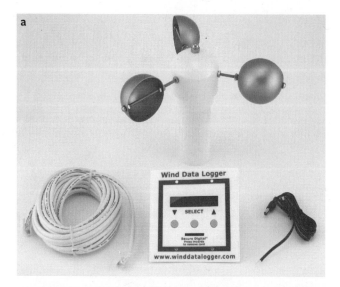

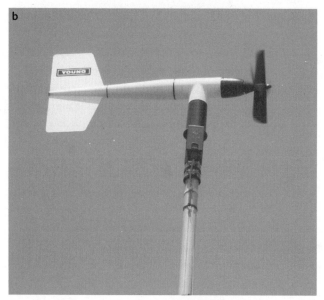

FIGURE 4.5. Anemometers. (*a*) Cup anemometer. (*b*) Propeller anemometers like this one are commonly used to measure wind speeds by the US Weather Service. Source: (*a*) APRS World, LLC. (*b*) Dan Chiras.

to a central hub (Figures 4.5a). The other version is a propeller anemometer. It consists of a miniature wind turbine (Figure 4.5b).

Anemometers and towers can be purchased online. Here's the rub, however: ideally, for best results, an anemometer needs to be mounted at the same height as the proposed wind turbine. That usually requires a tower of 100 to 120 feet (30 to 36 meters).

To assess wind speed at a site, you'll also need some way to record data, for example, a data logger. APRS World in Winona, Minnesota, sells an anemometer and a data logger. The data logger records speed and direction as well as time and date. Their data logger stores data on a removable 2-gigabyte SD (Secure Digital) card that can store many months' worth of data when measurements are recorded every 30 seconds. (Longer storage periods are possible with less frequent measurement.) At the end of each day, the data is stored in a file. The card is removed periodically and plugged into a computer. The data can be viewed in almost any spreadsheet program, graphed, and analyzed. Other data loggers are also available in the United States, Canada, and Europe and can be found online.

Although I don't recommend direct measurement for reasons that will be clear shortly, if you want to measure your wind resource and don't have the money to buy an anemometer, try contacting your state energy conservation office. They may have an anemometer and towers that they lend to individuals interested in assessing the wind resource at their sites. Even if your state offers this service, availability may be limited. The state, for instance, may only have four or five anemometers. If they're all tied up for a year or two, you will have to wait. In addition, most states provide short towers for their anemometer kits, the use of which will very likely provide you with useless data for sites with trees or other ground clutter. (Remember, you should be monitoring wind speeds at proposed tower height.)

Although direct measurement using a recording anemometer is the best way to determine average wind speed, doing so is not cheap if you have to purchase the equipment. An anemometer typically costs $400 or more. You'll also need to add the cost of an appropriately sized tower and guy cables to secure the tower. With tower costs and professional installation, direct measurement could easily cost $5,000 to $10,000, or more. As a result, Mick Sagrillo notes, "No one in the small wind industry puts up anemometers anymore."

Whatever you do, don't invest in a small rooftop or handheld anemometer. In the wind industry, they are merely toys. For accurate direct measurement, you'll need a durable anemometer on a very tall tower.

Airport or Weather Bureau Data

Another possible source of data on wind speed is a nearby airport or weather bureau. Airports and weather bureaus have been collecting data on wind speed regularly for ages. Unfortunately, this data has some serious shortcomings. One of them is that airport and weather bureau data is typically collected by anemometers on very short towers. It is then adjusted to the 30-foot standard for reporting using some simple math, presented in the textbox, "Estimating Wind Speed from a State Wind Map." You want data at tower height—100 to 120 feet or more!

While this approach sounds reasonable, it turns out to be a rather poor way of estimating wind speed. One reason is that many airport anemometers are installed in poor locations. For instance, they may be mounted next to a hangar or on the roof of a building. They may be located next to trees or in depressions where they are sheltered from the winds. In some cases, they are mounted near roadways where they are affected by car and truck traffic. Bear in mind that ground drag and turbulence at these sites greatly diminish the wind speed. Extrapolation of these artificially low values from anemometers that are 10 feet (3 meters) off the ground to the values expected on a 30-foot (10 meter) tower to a 120-foot tower (36 meters), therefore, generally underestimate the true wind speed at tower height. Although stronger winds might be present at the greater heights at which one would mount a wind turbine, you simply can't tell from the data.

The second problem with airport wind speed data is that airports are frequently sited in the least-windy locations to ensure safer takeoffs and landings. Runways at small airports may also be sheltered by trees to reduce surface winds. Therefore, wind speed measurements recorded at airports could underrepresent the wind potential of a nearby site. The same is true for cities where most people live and where weather bureau data is collected.

The third problem with airports is that airport personnel are not trained to monitor wind, and many sites have wind monitoring equipment that is not maintained well or is not working properly.

The fourth—and most critical problem—comes from trying to extrapolate the 30-foot data to the hub height at which a wind turbine should fly. Extrapolating wind speed up from one data point is iffy at best. This technique is little better than "wet finger prospecting."

I wouldn't recommend a wind turbine installation based on airport wind data. It's just too risky.

As Mick Sagrillo put it, airport data makes a lousy yardstick for determining wind speed at nearby sites. He adds, "Their average wind speeds are in all likelihood very low baseline numbers, really just a starting point for consideration. Virtually all wind generator sites I have seen have higher winds speeds by at least a mile per hour or two" above extrapolated wind speeds. If you don't think that two miles per hour is that much, remember that there is twice the available power in a ten mile-per-hour average wind speed than in an eight mile-per-hour average wind speed—that is, there's a 100 percent increase in available power from a 25 percent increase in wind speed!

Many times the disparity is three or four miles per hour.

Although airport and weather bureau data is widely available, we believe it should be used as a last resort, and only if you live very near the airport and the topography and ground clutter of your site is very similar to the airport's.

Far superior data is available on state wind resource maps and on the internet.

State Wind Maps

Because direct measurement is costly and time-consuming and airport data and weather bureau data are unreliable, professional small wind site assessors get the data they need from other sources. One of the best and easiest ways to assess the wind resources in a region is to consult a state wind map.

State wind maps list average annual wind speed in meters per second and miles per hour. Unfortunately, wind speed estimates on the map are typically reported at 165 and 195 feet (50 or 60 meters) above the surface of the ground.[1] Most residential wind turbines, however, should be mounted at about 100 to 120 feet (30 to 36 meters), sometimes as low as 80 feet (24 meters). Fortunately, a wind site assessor can use the data to extrapolate downward. For curious, mathematically inclined readers, the accompanying textbox explains how this is done.

Wind maps have come a long way in the past two to three decades, and most states have good ones. You can locate your state's wind map at a couple of different websites. Search for state wind maps—US Department of Energy. For Canadian readers, check out the *Canadian Wind Energy Atlas*. It's available online and provides a wealth of information including information on the direction of winds at various sites.

While wind maps are an excellent source of information for your area, they do have some limitations. One of them is resolution. In some areas, like the plains of western Kansas, wind maps show uniform wind speeds over large areas. If you live in one of those areas, the map will give you a pretty accurate idea of average wind speed. In other areas, however, the topography is much more complex. Wind speeds may vary considerably over short distances in such terrain, and the resolution of the maps isn't good enough to pinpoint an exact location. My former home in the foothills of the Rocky Mountains is a good example. The state wind map shows three different classes of winds in this area. Which one you're in, depends on where you live. A neighbor half a mile away in a valley may have very little wind, while a neighbor perched on top of one of the many mountains nearby will have a great deal of wind. While this seems like common sense to someone experienced in siting wind turbines, it often goes unnoticed by newcomers to wind.

In assessing wind resources, location is everything. Local topographic features can confuse matters considerably. So, just because a map shows that wind speeds in your area are sufficient, it doesn't mean they are. Hills, cliffs, forests, and buildings can deflect winds, greatly reducing wind flow at downwind sites. Some types of hills and cliffs can even magnify winds (Figure 4.6). So, the more complex the terrain, the less accurate the wind maps are.

Using data from wind maps, it is possible to make an initial, approximate determination regarding the suitability of a site. However, it's always a good idea to hire

FIGURE 4.6. Effect of Topography on Wind Speed. Hills can dramatically increase wind speed. Placement of a tower on the top of the hill could result in a significant increase in the power output of a turbine. Placement of a turbine at the base of a hill could result in much lower output. Credit: Anil Rao.

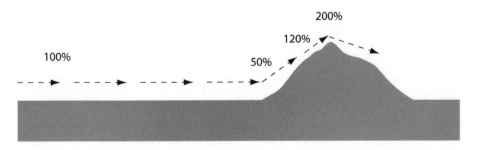

Estimating Wind Speed from a State Wind Map

State wind maps provide annual average wind speed data, but it is measured considerably higher (165 to 195 feet, or 50 to 60 meters) than most residential wind towers (80 to 160 feet, or 24 to 48 meters). Fortunately, average wind speed data from state wind maps can be adjusted using a relatively simple equation: $V = (H/H_0)^\alpha V_0$.

In this equation, H is the height of the tower on which the turbine will be mounted. H_0 is the height at which the average wind speed data is reported. V is the speed of the wind at the desired tower height, which is unknown. V_0 is the average annual wind speed recorded on the state wind map. The Greek symbol α, or Alpha, is the wind shear coefficient. The shear coefficient is a measure of the change in wind speed with height above the ground caused by surface texture or roughness, which is caused by ground clutter such as trees and buildings. It is used to make adjustments in wind speed at different heights based on the "texture" or "roughness" of the ground surface.

As noted in Chapter 2, the rate at which wind speeds increase with height varies based on the ground surface, notably the type of vegetation and the topography. The greatest increases occur over rough terrain, that is, terrain with numerous obstacles, such as trees and shrubs. The rate of increase is the lowest over smooth terrain, for example, a short-grass prairie or a meadow. Table 4.4, lists typical shear coefficients.

To estimate wind speed at a proposed tower height, simply fill in the numbers and do the math. You will need a calculator with the exponent function, which all but the simplest ones have. To see how this works, suppose that average wind speed from the state wind map at 198 feet at a rural site with scattered trees and building is 15 miles per hour. The wind shear coefficient is 0.24, according to Table 4.4.

To calculate the average wind speed at 120 feet, we begin with the equation:

$$V = (H/H_0)^\alpha V_0$$

To solve the equation, plug in the values.

$$V = (120/198)^{.24} \times 15 = (0.61)^{.24} \times 15 = 13.3 \text{ mph}$$

The average wind speed at the proposed height is about 13.3 miles per hour. Professional wind site assessors round down to be conservative. They'd use 13 miles per hour to determine the electrical production of a wind turbine.

When using this equation, keep in mind that the height measurements must be in the same units.[2] Both must be in feet or both must be in meters. Similarly, the speed data must be in consistent units, either miles per hour or meters per second. Don't mix units or the results will be wrong.

To calculate exponents, you can use a calculator. Google also provides a very convenient calculator. Simply type in the equation in the search window and hit the search button.

Table 4.4. Typical Wind Shear Coefficients

Surface Characteristics	Shear Coefficient
Lake or ocean, water or ice	0.10
Short grass or tilled ground	0.14
Level country, foot high grass, occasional tree	0.16
Tall row crops, hedges, short fence rows	0.20
Hilly country with open ground	0.20
Few trees, occasional buildings	0.22
Many scattered trees, more buildings	0.24
Wooded country, small town	0.28
Suburbs	0.30
Urban areas	0.40

a professional to do an on-site analysis. He or she will study the topography, vegetative cover, and ground clutter to arrive at a more accurate estimate of average annual wind speed at your site. But how much wind do you need to make a system worth the investment?

Mike Bergey states that a stand-alone (off-grid) wind energy system makes sense if you're in an area with an annual average wind speed of eight miles per hour (3.6 meters per second) or higher at hub height. It should, he contends, be economically feasible. Mick Sagrillo agrees, and says this is especially true for those whose off-grid systems incorporate PV. This makes a lot of sites in the United States candidates for off-grid homes powered by wind systems.

Remember, though, off-grid homes tend to use very little electrical energy—one-tenth to one-twelfth as much energy as a grid-tied home. Off-gridders dry clothes on clotheslines, cook on propane or natural gas, install only super-energy-efficient appliances and lighting, and work hard to save energy. They are not your typical US or Canadian energy guzzler.

While an 8-mile-per-hour average wind speed may suffice for an off-grid system, according to Bergey, grid-connected systems typically require a higher average annual wind speed of about 10 miles per hour (4.5 meters per second) to be economical. Again, this speed is at the hub height of the turbine. I advise caution when viewing general recommendations such as these. With the higher cost of wind turbines these days, higher wind speeds might be necessary to make a wind system economical—perhaps 12 miles-per-hour (5.4 meters per second). All in all, it is best determine the wind speed at your site; determine which turbine you need to install to meet your needs; determine the cost of the system, including the tower; and run calculations to determine the economics of this proposed installation. I'll show you some ways to do this at the end of this chapter.

Online Databanks

While state wind maps are an excellent source of information, they have their shortcomings, as just noted. Because of this, some professional small wind site assessors perform small wind site analyses using an amazing database developed by NASA: the website Surface Meteorology and Solar Energy.

Surface Meteorology and Solar Energy provides a wealth of data on wind and solar energy. In the section on wind energy, it presents tables that show both the monthly and annual average wind speed and the direction of the wind. The site contains data for the entire world. To use this site, you will need to set up an account. It's free and it only takes a few seconds. The accompanying textbox provides a few helpful hints that will make your visit to the site more productive.

Tips on Using Surface Meteorology and Solar Energy Site

After signing up for and logging onto the NASA site, Surface Meteorology and Solar Energy, click on "Meteorology and Solar Energy." Next click on "Enter the Latitude and Longitude." Enter this data for your site.

Fortunately, there are numerous websites that will assist you in finding your latitude and longitude. Some allow you to pinpoint your city or town on a map. I recommend the less common sites that enable you to find the latitude and longitude of your site by entering your street address. Search for "find latitude and longitude by address" to find these sites.

When using the NASA site, confusion may occur when you enter your latitude and longitude. Latitudes and longitudes may be entered as decimal degrees and minutes separated by decimal points (33.5) or as degrees and minutes separated by a space (33 50).

As I hope you remember from middle school geography, the lines of latitude run parallel to the equator—from the equator to each of the poles. The equator is assigned a latitude of zero degrees. The North and South Poles are 90 degrees. Lines of latitude may be either north or south, depending on whether you are north or south of the equator. To avoid confusion, south latitudes are assigned a minus sign.

As you may also recall, the lines of longitudes run from the North to the South Poles. By convention, zero degrees longitude (the prime meridian) passes through the Royal Observatory in Greenwich, England. Locations east of the prime meridian are assigned a positive number. Latitudes west of the prime meridian are assigned a minus sign. (Lines of longitude run from 0 to 180 degrees.)

OK, enough geography. Once you have entered the latitude and longitude, click on "Submit." You'll now be presented with a long list of data from which to choose.

Scroll down to the first box on wind, entitled "Meteorology (Wind)." In this box, you can click on one of several options. I usually click on "Wind direction at 50 meters" to determine the predominant wind flow at a site. This number will be expressed in degrees, and is the direction the wind is coming from. Zero degrees is north. Ninety degrees indicates wind from the east; 180 degrees indicates wind from the south, while 270 degrees indicates wind from the west.

In the box below, click on "Wind Speed at Any Height and Vegetation Type."

Now scroll through the list to select the vegetation type that best represents the site you want to assess. This will enter the shear factor, or Alpha, discussed earlier in this chapter.

Now enter the height of the proposed wind tower in meters. To convert feet to meters, divide feet by 3.28.

Now scroll to the bottom of the page and click on "Submit." You'll be presented with the data you requested, in this case the predominant wind direction and average wind speed data both by month and by year. Note that the data is presented in meters per second. To convert meters per second to miles per hour, multiply by 2.237. Data tables can be cut and pasted into documents for reports or for your own use later on.

Like wind maps, NASA data provides good information, but you may still need to assess your site very carefully, especially in complex topography. If your site is in a valley or hilly terrain, for instance, the average wind speed determined by the website may not represent the average wind speed at your site. If the proposed turbine is downwind from numerous buildings or a forest, the wind speed at the wind generator may be lower that the satellite data suggests.

By all means, hire a professional wind site assessor to analyze your site and make recommendations for tower/turbine placement and minimum acceptable tower height. You may also want to ask a professional installer to estimate the wind speed at your site, although not all installers are as knowledgeable about wind and wind site assessment as certified wind site assessors. Some installers may be more interested in selling and installing a turbine and tower than in giving you an objective wind site assessment.

You can also obtain a wind site analysis online from wind turbine manufacturers like Bergey WindPower and Xzeres Wind. (Bear in mind, if you enter your data, however, you'll very likely be contacted by a salesperson from these manufacturers.)

Flagging and Other Factors

Even if a state wind map or the NASA data suggests the presence of a good resource on a site, it is important—no, essential—that you or your site assessor study other factors such as the topography, vegetative cover, the proximity of the wind turbine to trees, hills, cliffs, and buildings, all of which could decrease—or, in some cases, increase—wind speeds. You or your site assessor should also take into account local wind patterns like mountain-valley winds and offshore and onshore winds that might boost your wind system's output.

In some areas, vegetation provides an important clue as to how windy a site is. Figure 4.7 shows a tree in the windswept foothills of the Rockies. This tree exhibits a phenomenon known as flagging. What is flagging?

As readers know, a flag blowing in the wind blows to one side of the pole. Strong winds produce a similar but permanent visual effect in trees, especially coniferous trees; this is called flagging. When flagging occurs, the branches of a tree appear to stream downwind. Persistent strong winds do not cause the branches of a tree to bend around and "fly" on one side of the trunk. Rather, they damage or stunt the growth of upwind branches, so it appears as if the downwind branches are streaming downwind like a flag. In really strong winds, the upwind branches may be stripped away entirely.

FIGURE 4.7. Flagging. This photo shows a windy site in the foothills of the Rockies in Colorado. The pine tree in the distance exhibits moderate flagging, indicating an average annual wind speed of 11 to 13 miles per hour. The pine tree in the middle exhibits complete flagging, indicating 13 to 16 mile-per-hour average annual wind speed. Both trees are on the same property, illustrating how wind can vary within a short distance. Credit: Dan Chiras.

Flagging can be used to approximate the average wind speed at some sites, as shown in Figure 4.8. This scale is known as the Griggs-Putnam Index. According to this index, slight deformity, known as *brushing and slight flagging*, indicates a probable mean annual wind speed in the range of seven to nine miles per hour. Slightly more flagging, cleverly referred to as *slight flagging*, indicates an approximate average annual wind speed of 9 to 11 miles per hour. Moderate flagging indicates an average annual wind speed of 11 to 13 miles per hour, and so on. On the extreme upper end, the trees are bent to the ground, a phenomenon called *carpeting*.

Flagging is typically observed in single trees—that is, a tree growing alone in a field isolated from other trees and obstacles. Trees in forests or groves are not prone to flagging except at their tops, because they shelter one another.

Flagging can be used to verify data from NASA or wind maps. If the wind map or NASA data indicates that a site has an average wind speed of 15 miles per hour and the trees exhibit complete flagging, you can feel confident that the data is correct.

Although flagging helps verify other data, it is important to note that the absence of flagging does not necessarily indicate a lack of wind. Some trees aren't very susceptible to flagging. Most deciduous trees, for example, are less prone to flagging than coniferous trees. That said, if you observe carefully, you can find signs of heavy winds in the way deciduous trees grow. Take a look at Figure 4.9. It shows a tree along I-70 in western Kansas that is bent away from the wind thanks to the persistent southwesterly winds that sweep across the Plains.

Site assessors and homeowners can also look for other telltale signs of strong winds on a site. If neighbors or local governments use snow fences to prevent snow from drifting over driveways and roads, it's a good indication of strong winter winds (Figure 4.10). Clouds of dust and dirty snow are signs of soil erosion by wind, a good sign for wind site assessors (a bad sign for farmers). The use of windbreaks, trees planted upwind from homes and farmsteads to protect them from winds, is another

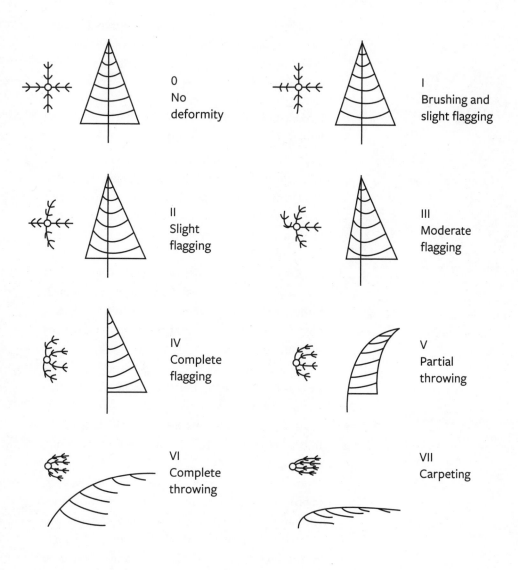

Griggs-Putnam Index of Deformity

Index	I	II	III	IV	V	VI	VII
Wind Speed (mph)	7–9	9–11	11–13	13–16	15–18	16–21	22+

FIGURE 4.8. Griggs-Putnam Index of Deformity. Credit: Paul Gipe.

FIGURE 4.9. Blowing in the Wind. Strong, persistent winds can cause trees to bend over like this one at a roadside stop in western Kansas. Credit: Dan Chiras.

FIGURE 4.10. Living Wind Fence. Trees (background) planted along Interstate 70 in windy western Kansas prevent snow from drifting across roadways. Notice the humanmade snow fence in the foreground to provide added protection. Humanmade and living snow fences are a good indication you've got a windy site. Credit: Dan Chiras.

good sign that there's wind to be reckoned with. Tattered flags flying on flagpoles tell the same story. Telephone and electric poles that tilt at an angle may convey a similar message.

How Much Electricity Will a Wind Turbine Produce?

Once you've determined the average wind speed at a site, it is time to determine how much electricity a wind generator could produce at the proposed tower height—and therefore whether it can meet all your needs or what percentage of your needs it will satisfy. This step is fairly easy.

Table 4.5 shows a list of numerous small wind turbines and the estimated annual output of each turbine (in kilowatt-hours) at seven different average wind speeds. To see how this table is used, consider an example.

Let's assume that the average wind speed at a site (at hub height) is 12 miles per hour. Let's also assume that your load analysis, after efficiency measures have been implemented, indicates you'll need 16,500 kWh per year. In the 12 mile-per-hour column, you'll discover one wind turbine that will meet your needs: the Bergey Excel 10, a turbine that has been in production for over three decades. (It also comes with an unprecedented 10-year warranty.) You will also notice another turbine that produces a little more than you'll need, the Xzeres 442SR. It produces about 20,000 kWh, according to the manufacturer.

Annual energy outputs like those shown in Table 4.5 are used to determine which turbine you need. If you want to find more, be sure to check out *Home Power* magazine. They publish an annual "Wind Turbine Buyer's Guide" that typically lists the output of 18 or so small wind turbines. Check out the May/June issue of 2015, as an example.

Table 4.5. Estimated Annual Energy Output in kWh of Selected Small Wind Turbines According to Manufacturers

Turbine	8 mph	9 mph	10 mph	11 mph	12 mph	13 mph	14 mph
Bergey Excel 1	420	610	840	1,180	1,400	1,710	2,040
Luminous Whisper 200	871	1,231	1,629	2,048	2,473	2,890	3,291
Pika T 701	700	1,250	1,800	2,350	2,900	3,500	4,100
Xzeres Skystream 3.7	989	1,740	2,576	3,282	4,115	4,962	5,814
Luminous Whisper 500	2,309	3,286	4,386	5,572	6,803	8,042	9,256
Luminous Windistar 4500	2,724	4,800	6,900	7,800	9,600	12,000	14,400
Bergey Excel 6	3,963	5,582	7,470	9,536	11,667	13,850	16,325
Ventera VT-10	5,037	7,218	8,957	11,625	13,924	17,599	20,836
Bergey Excel 10	4,924	7,124	9,850	13,026	16,499	20,248	24,712
Bergey Excel R	4,549	6,723	9,292	12,114	15,008	17,922	21,125
Xzeres 442SR	4,990	8,583	12,630	16,017	19,958	23,984	27,997

You can also obtain annual energy output data directly from the websites of wind turbine manufacturers, though this can be quite time-consuming. While this data is useful, some wind installers argue that manufacturers overstate the electric production of their turbines. As a result, you may want to derate their estimated outputs by 10 to 20 percent—just to be conservative. (The data in *Home Power* magazine is not derated.)

Once you've found the wind turbines that meet your need at the proposed tower height, it's time to consider the cost.

Does a Wind System Make Economic Sense?

At least three options are available when it comes to analyzing the economic cost and benefits of a wind turbine: (1) a comparison of the cost of electricity from the wind turbine with conventional power or some other renewable energy technology, (2) an estimate of simple return on investment, and (3) a more sophisticated economic analysis tool known as discounting. I'll present an overview of each method in this chapter.

Cost of Electricity Comparison

One of the simplest ways of analyzing the economic performance of a wind energy system is to compare the cost of electricity from the wind system to the cost of electricity from a conventional source—notably, the local utility—or some other renewable energy technology you are considering. This is a five-step process, three of which we've already discussed.

After identifying a wind turbine that produces a sufficient amount of electricity to meet your needs, divide the cost of the system by the total output over the lifetime of the turbine. A well-built turbine should last 20 to 30 years. A less sturdy unit may not make it 10 years.

Let's consider an example. If the Bergey Excel 10 produces 16,499 kWh per year (based on manufacturer's estimate) at a site with a 12 mile-per-hour average wind speed at hub height, it will produce 329,980 kWh over its 20-year lifetime. The wind turbine itself sells for $32,000, not including tax. A good 100-foot guyed lattice tower will cost another $14,000. Installation could run around $20,000 give or take a little. Total cost, then, is about $66,000. (Costs vary depending on travel distance to your site, accessibility by crane, labor costs, and other factors.) If the homeowner receives a 30% Federal Tax Credit, the cost would be $46,200. The electricity the turbine will produce over its lifetime will run you 14 cents per kilowatt-hour. You calculate that by dividing the cost, $46,200, by the 20-year production, 329,980 kWh.

Is this a bargain?

It depends.

If the alternative source costs 17 to 30 cents per kilowatt-hour (or more), this system turns out to be a good investment. High rates like these are not common in the center of the United States, but they certainly are encountered on the east and west coasts. They are also common prices on islands, like Hawaii.

This system could make economic sense from the get-go, but only if you don't mind prepaying your electrical bill—laying down $66,000 all at once. (You'll get the 30% federal tax credit of $19,800 back in taxes. Homeowners can carry the tax credit for two years.)

If you are installing this wind turbine on a business, you not only receive the 30% Federal Tax Credit, which you can carry for 20 years, but you can also deduct 50% of the expense of the system (including its installation) in year one. The remainder is depreciated over the next 4 years. Depending on your tax bracket, this could reduce the initial cost by 15 to 30 percent, maybe more.

If this system is installed to power a farm or ranch, IRS ruling 179 allows an owner to deduct the full expense from his or her income. In other words, it is a complete write off. Rural businesses can also receive 25% grants from the US Department of Agriculture. (They are called REAP grants, which stands for Rural Energy for America Program.)

To see how these incentives help, let's take the same example from above and assume that a rural business, say a small rural factory, installs a Bergey Excel 10. The initial cost is $66,000. The business receives a 25% USDA REAP grant. That drives the cost down to $49,500. (The money is deposited directly into the recipient's checking account after the system is up and running.) The business also receives a 30% Federal Tax Credit. That drives the price down to $34,650. (The FTC is calculated after the REAP grant.) The business can also depreciate the cost of the equipment over 5 years. Let's suppose that saves the business $8,000 in taxes. The cost is now $26,650. What's the cost of electricity?

This time, you calculate the cost per kilowatt-hour of electricity by dividing the cost, $26,650 by the 20-year production, 329,980 kWh. Now the cost of electricity has dropped to 8 cents per kilowatt-hour. That's the cost in many rural areas in the United States—especially in the heartland of the country, where service is typically provided by rural electric co-ops. If the business is serviced by an investor-owned utility, the big utilities that provide energy to cities and outlying rural areas, costs are very likely 10 to 12 cents per kilowatt-hour.

The cost of electricity calculations does not include several important factors. For example, it does not include insurance or maintenance costs. And, lest we forget, you need to have access to the money you need to pay the installer for your system. Or you'll need to be able to borrow it.

To make a fair analysis, you should factor in the cost of lost economic opportunities, the money you could have made on that money had you invested it in something else.

The comparison to local utility rates, however, does not take into account that the cost of electricity from the utility is also bound to increase. In fact, it has been rising nationwide at a rate of 4.4 percent per year for the past four decades—and could increase more rapidly as energy prices rise. Would the rise in electrical costs offset lost income or interest paid on a loan? Probably. But probably not the cost of insurance (property and liability insurance, discussed in Chapter 10) and increased property taxes, if any, resulting from the installation of the turbine. (Some states waive property taxes on renewable energy systems.)

When calculating the cost of electricity from a wind system, don't forget to subtract financial incentives from state and local government or local utilities from the initial cost. These incentives can be quite substantial. To learn more about incentives in your state, log on to the Database of State Incentives for Renewables and Efficiency, dsireusa.org. Study them carefully. Ask local installers about state and local incentives as well.

Here's another thing to consider if you are building a new home in a rural area. If you are considering going off-grid, it may be cheaper to install a wind or solar system than to connect to the electric utility. If your home is more than a few tenths of a mile from existing electric lines, you could face a huge bill to connect to the grid—as much as $20,000. If your home is located even farther away, the cost could be higher. A half-mile line could cost as much as $20,000 to $50,000. Remember, the cost of connecting to the electric grid only covers the cost of the installation of poles and electric lines. It does not include a single kilowatt-hour of electricity. As a result, if you're building a new home that is more than a few tenths of a mile from a power line and the utility company requires customers to pay for line extension, an off-grid wind system that costs $46,000 may be well worth the investment. (But don't forget to add in the cost of other equipment, such as a battery bank.)

Comparing the cost of electricity from a wind energy system to the cost from conventional power (or another source of renewable energy) is a very simple way to assess economic feasibility. To perform a detailed financial analysis, however, it is best to include *all* of the costs that you can reasonably predict and estimate. If you need help, you may want to hire an accountant to run the numbers for you. Or you can use the spreadsheet that I'll introduce shortly. It will handle all of this for you, and allow you to compare more accurately the cost of a wind energy system over the lifetime of the system to the cost of utility power.

Calculating Simple Return on Investment

Another relatively simple way of determining the cost effectiveness of a renewable energy system is to determine the simple return on investment (ROI). Return on investment is, as its name implies, the rate of return expressed as a percentage of an investment.

ROI can be calculated by dividing the annual value of electricity generated by a wind system by the cost of the system. Let's stick with our previous example. If this wind system produces 16,499 kWh per year, and you are paying 12 cents per kilowatt-hour for electricity you purchase from your local utility, the system will produce $1,980 worth of electricity each year. If the system cost $46,200 to install, the simple return on investment is $1,980 divided by $46,200, or 4.3 percent. Not terribly good, but better than savings accounts at banks, which are paying 0.125 percent interest, or the best online savings accounts that pay a measly 0.95 percent. This simple ROI is improved if you factor in all of the feel-good variables, like producing your own power from a clean renewable energy source and watching your wind turbine on blustery days generating power.

If the turbine cost is $26,650 for a business (for reasons outlined above), the simple ROI would be $1,980 divided by $26,650, or 7.4 percent.

Like the previous method, simple return on investment does not take into account a number of economic factors that could influence one's decision. For instance, it does not include interest payments on loans required to purchase the system or lost interest if the turbine and tower are paid from an interest-bearing account. It does not take into account insurance costs or property taxes. All of these factors decrease the value of the investment.

On the other side of the ledger, this simple calculation does not take into account factors that increase the value of the investment, for example, the potential increase in the cost of electricity from the local utility. This would make wind-generated electricity more valuable. Moreover, this calculation does not take into account the fact that money saved on the utility bill is tax-free income to you.

This analysis also does not account for the potential use of electricity as a transportation fuel, for example, in electric cars or plug-in hybrids. While electricity is typically used to power lights, appliances, and electronics in a home and business, it can also be used to power vehicles.

My Nissan Leaf travels 4.4 miles on a single kilowatt-hour of electricity. To travel 30 miles, it takes 6.8 kWh. We pay 8 cents per kilowatt-hour at our rural location. As a result, it will cost me 55 cents to drive 30 miles, if I buy the electricity. If I drove a car that got 30 miles per gallon, it would cost me $2.00 at current fuel prices (August 2016). The 6.8 kWh of electricity would therefore displace $2.00 worth of gas. The value of each kilowatt-hour would be 29 cents. If the price of gas rises to $3.00, the value of my wind-generated kilowatt-hour would be 44 cents. If it rises to $4.00 per gallon, the value of my wind energy would be 59 cents per kilowatt-hour.

My electric car is used for local trips, so we leave the Prius at home. We drive the Prius about 600 miles less each month as a result of our heavy dependence on the Nissan Leaf. Because the Prius gets 50 miles per gallon, the value of our solar- and wind-generated electricity is a bit lower than the previous calculations. To drive the Leaf 50 miles, I'd need 11.4 kWh of electricity. With gas at $2.00 per gallon, the electricity's value is still high—a little over 17 cents per kilowatt-hour. If gas prices rise to $3.00, the value of the electricity climbs to 26 cents. If $4.00 becomes commonplace again, the value of my electricity would be 35 cents. Not bad, eh?

Simple return on investment is a great way of evaluating the economic performance of a renewable energy system. There is one thing, though, I should note. Unlike a bank account in which you invest money and receive interest as your return on investment, when it comes to a wind or solar system, the principal does not grow, it decreases in value. That's why I label this a "simple" return on investment.

Simple ROI is valuable, nonetheless, to my customers, as it helps them understand what they are receiving in return for their investment in renewable energy. And, it's light years ahead of the black sheep of economic tools: payback. For reasons that will be clear shortly, I and other professionals avoid payback discussions at all costs when talking to clients or in the classes and workshops we teach.

Why?

As most readers know, payback is a term that gained popularity in the 1970s. It is used for energy conservation measures and renewable energy systems. More recently, payback has been applied to energy-efficient hybrid vehicles and electric cars. Payback is, quite simply, how long it takes a system or energy conservation measure in years to pay back its cost through the savings.

For a wind energy system, payback can be determined by dividing the cost of the system by the anticipated annual savings. If the $46,200 wind energy system we've been looking at produces 13,499 kilowatt-hours per year and grid power costs you 12 cents per kilowatt-hour, the annual savings of $1,980 yield a payback of 23 years ($46,200 divided by $1,980). In other words, it will take the savings from your system 23 years to pay off the cost, ignoring the maintenance and repair costs. From that point on, the system produces electricity free of charge. If the system cost a business, $26,650, the payback would be 13.5 years.

While 23-year and 13.5-year payback seem ridiculously long, remember, the return on investment, calculated earlier, was actually 4.3 percent in the first case and 7.4 percent, respectively, both fairly decent returns on investment. Interestingly, while the 7.4 percent return on investment seems fairly decent, the 13.5-year payback, which represents the same return, seems prohibitively long.

Simple payback is pretty straightforward and fairly easy for most of us to understand, but it has very serious drawbacks. The most important is that it is seriously misleading, as the previous example illustrates. While you might love a 10 percent return on your investment, a 10-year payback seems awfully long when, in fact, they're exactly the same, assuming we ignore opportunity costs, inflation, and other factors when calculating both payback and return on investment.

Simple payback and simple return on investment are closely related metrics. Mathematically speaking, return on investment is the reciprocal of payback. That is, ROI = 1/Payback. As just noted, a wind system with a 10-year payback has a 10 percent return on investment (ROI = 1/10).

One final word on the subject: it's worth noting that we rarely use payback, or ROI for that matter, when considering purchases in our lives. Do we calculate the payback of our bass boat costing $25,000 or a zero-radius lawn mower that costs $4,000?

Discounting and Net Present Value: Comparing Discounted Costs

The best way to determine whether an investment in wind energy makes economic sense is to compare the cost of the system to the cost of electricity it will displace. This analysis, unlike the others we have presented in this chapter, takes into account maintenance costs and the rising cost of grid power. It also takes into account the time value of money—the fact that a dollar today is worth less tomorrow, and even less a few years from now. These calculations incorporate a factor economists call a discount rate that allows them to factor in the declining value of money, as I'll explain shortly.

To make life easier, this economic analysis can be performed by using a spreadsheet like the one shown in Table 4.6 provided by renewable energy economist John Richter of the Sustainable Energy Education Institute in Michigan. This spreadsheet is available for your use on my website, evergreeninstitute.org, courtesy of John. I recommend that you use it.

To understand how this technique is used, consider an example. I'll use the same turbine we've been studying. Table 4.6 shows the economic analysis for a 10 kilowatt-hour wind turbine that costs $46,200 to install. Grid power costs 12 cents per kilowatt-hour. As you can see by examining Table 4.6, the spreadsheet contains several columns. The first column on the left is years—how far out you want to run the calculation. This analysis was run for a period of 20 years, the lifetime of a good wind system—though, you may get more years out of a system if you perform maintenance and install it in a low-turbulence site (on a tall tower).

The next column is the discount factor. That's the decline in the value of the dollar. We'll use 3 percent per year.

The next column (under the category "Buy Utility Electricity") shows the cost of electricity from the local power company—that is, how much you would be paying each year for electricity purchased from the local utility—taking into account rising fuel costs (4.4 percent annual increase). The top number is the amount of electricity your system will produce multiplied by the cost of electricity from the utility. In our case, that's $1,980 in year 1. Notice that in year 2, you will pay $2,067 for the same amount of electricity because of the 4.4 percent increase in price. In year 3, you will pay more, and so on. The bottom figure in this column is the total cost of electricity—$61,469. That's how much money you will spend over 20 years buying 16,499 kWh of electricity from the local utility, taking into account an average increase in cost of 4.4 percent. (It is the amount you would get if you added up all the checks you wrote over 20 years.)

The next column under "Buy Utility Electricity" is the discounted cost of electricity from the utility—the cost of electricity adjusted for the increase in the cost

Table 4.6. Economic Analysis of the Cost of a Wind Energy System

Year	Discount Factor	Buy Utility Electricity		Proposed Wind System	
		Cost	Discounted Cost	Cost	Discounted Cost
Rate:	3.0%	4.4%			
0	1.000	$0	$0	$46,200	$46,200
1	0.971	$1,980	$1,922	$0	$0
2	0.943	$2,067	$1,948	$100	$94
3	0.915	$2,158	$1,975	$0	$0
4	0.888	$2,253	$2,002	$100	$89
5	0.863	$2,352	$2,029	$0	$0
6	0.837	$2,456	$2,057	$100	$84
7	0.813	$2,564	$2,085	$0	$0
8	0.789	$2,677	$2,113	$100	$79
9	0.766	$2,794	$2,142	$0	$0
10	0.744	$2,917	$2,171	$100	$74
11	0.722	$3,046	$2,200	$0	$0
12	0.701	$3,180	$2,230	$100	$70
13	0.681	$3,319	$2,260	$0	$0
14	0.661	$3,466	$2,291	$100	$66
15	0.642	$3,618	$2,322	$0	$0
16	0.623	$3,777	$2,354	$100	$62
17	0.605	$3,943	$2,386	$0	$0
18	0.587	$4,117	$2,418	$100	$59
19	0.570	$4,298	$2,451	$0	$0
20	0.554	$4,487	$2,484	$100	$55
Total		$61,469	$43,840	$47,200	$46,933

of electricity *and* the decline in the value of money. In year 2, for instance, you will shell out $2,067 but that's only $1,948 in year 1 dollars. At the end of the 20-year period, the discounted cost of electricity—the cost in year 1 dollars—is $43,840, shown in column four. Even though your checks added up to $61,469, the present value, the value in year 1 dollars, is only $43,840.

This leads us to the final steps. In the fifth column of the spreadsheet, I entered the cost of the wind energy system—$46,200. This figure should include all of the costs described earlier in the chapter minus financial incentives you will receive. Notice in the example I've included a $100 maintenance expense every other year. As a result, the actual outlay at the end of 20 years will be $47,200.

The next and final column of the spread sheet is the discounted cost of the wind

Table 4.7. Economic Analysis of a Wind System Installation

Year	Discount Factor	Buy Utility Electricity		Proposed Wind System	
		Cost	Discounted Cost	Cost	Discounted Cost
Rate:	3.0%	4.4%			
0	1.000	$0	$0	$26,650	$26,650
1	0.971	$1,980	$1,922	$0	$0
2	0.943	$2,067	$1,948	$100	$94
3	0.915	$2,158	$1,975	$0	$0
4	0.888	$2,253	$2,002	$100	$89
5	0.863	$2,352	$2,029	$0	$0
6	0.837	$2,456	$2,057	$100	$84
7	0.813	$2,564	$2,085	$0	$0
8	0.789	$2,677	$2,113	$100	$79
9	0.766	$2,794	$2,142	$0	$0
10	0.744	$2,917	$2,171	$100	$74
11	0.722	$3,046	$2,200	$0	$0
12	0.701	$3,180	$2,230	$100	$70
13	0.681	$3,319	$2,260	$0	$0
14	0.661	$3,466	$2,291	$100	$66
15	0.642	$3,618	$2,322	$0	$0
16	0.623	$3,777	$2,354	$100	$62
17	0.605	$3,943	$2,386	$0	$0
18	0.587	$4,117	$2,418	$100	$59
19	0.570	$4,298	$2,451	$0	$0
20	0.554	$4,487	$2,484	$100	$55
Total		$61,469	$43,840	$27,650	$27,383

turbine. Like the discounted cost of electricity, it is the value of your expenditure in year 1 dollars. That is, it takes into account loss in the value of the dollar due to inflation.

The final step is a simple comparison of costs. Compare the discounted cost (present value) of electricity from a wind turbine to the discounted cost (present value) of electricity from the utility. In this example, the present value is $46,933, a bit higher than the present value of electricity, $43,840.

If the wind system is cheaper, it makes economic sense. If it costs more, it doesn't. If the differential is small—as it is here—you may want to go for it anyway.

Table 4.7, shows a similar analysis for the same system installed on a business, using a 4.4 percent increase in the cost of electricity from the local utility. In this

Table 4.8. Economic Analysis of a Wind Energy System with 10% increase in Electrical Rates

Year	Discount Factor	Buy Utility Electricity		Proposed Wind System	
		Cost	Discounted Cost	Cost	Discounted Cost
Rate:	3.0%	10.0%			
0	1.000	$0	$0	$46,200	$46,200
1	0.971	$1,980	$1,922	$0	$0
2	0.943	$2,178	$2,053	$100	$94
3	0.915	$2,396	$2,192	$0	$0
4	0.888	$2,635	$2,342	$100	$89
5	0.863	$2,899	$2,501	$0	$0
6	0.837	$3,189	$2,671	$100	$84
7	0.813	$3,508	$2,852	$0	$0
8	0.789	$3,858	$3,046	$100	$79
9	0.766	$4,244	$3,253	$0	$0
10	0.744	$4,669	$3,474	$100	$74
11	0.722	$5,136	$3,710	$0	$0
12	0.701	$5,649	$3,962	$100	$70
13	0.681	$6,214	$4,231	$0	$0
14	0.661	$6,835	$4,519	$100	$66
15	0.642	$7,519	$4,826	$0	$0
16	0.623	$8,271	$5,154	$100	$62
17	0.605	$9,098	$5,504	$0	$0
18	0.587	$10,008	$5,879	$100	$59
19	0.570	$11,009	$6,278	$0	$0
20	0.554	$12,109	$6,705	$100	$55
Total		$113,404	$77,074	$47,200	$46,933

instance, you can see that the discounted cost of the system ($27,383)—its cost in today's dollars—is considerably less than the discounted cost of electricity from the grid ($43,840).

While these analyses are based on an increase in utility electricity of a little over 4.4 percent per year, many areas are experiencing even greater increases. In St. Louis, for example, the cost of electricity has increased about 10 percent a year for the past seven years. Table 4.8 shows the effect that has on the present value of electricity vs. a residential wind energy system. As you can see, the economic prospects of the wind system change dramatically with rising energy prices. It's not a bad idea to invest now, if you are concerned about rising prices.

After determining the cost of electricity, calculating a simple return on investment, and comparing the present value of wind vs. utility power, you will be equipped with the information needed to make a rational decision about wind power. Don't forget, however, that there are other mitigating factors. As John Richter pointed out at a workshop at the 2006 Michigan Energy Fair sponsored by the Great Lakes Renewable Energy Association, "Even though a system may not make perfect sense from an economic standpoint, it is your money. You can spend it how you see fit." You may, for instance, purchase a system for peace of mind—knowing you will be free from utility power and not subject to rising fuel costs. Being independent of the power company is often a compelling reason for independent-minded folks. Creating an opportunity to sell power to the local utility may be worth the investment. Or, you may find the personal satisfaction of generating power from a clean, renewable resources sufficient. Just having a fancy new toy to play with may also be worth a slightly higher electrical cost.

Putting It All Together

Now that you have seen how one determines whether a wind energy system makes economic sense, it should be clear that blanket statements about "qualifying" wind speeds—wind speeds that make a wind energy system worthwhile—are only general guidelines. Economics is where the rubber meets the road in many cases. Comparing wind against the "competition," calculating the return on investment, or comparing strategies based on present value, gives us a more realistic view of the economic feasibility of wind energy at a site.

As I have already pointed out, economics is not the only metric by which we judge our decisions. Energy independence, environmental values, reliability, the "cool" factor, and the "fun value" of a hobby for the renewable energy gear heads are among the many factors that also play prominently in our decisions to invest in renewable energy. They are all perfectly valid motivations for installing a renewable energy system, and they should never be forgotten.

As Ian Woofenden points out, "We have certain values by which we strive to live, and we pay good money for things we value." People often invest in renewable energy because they want to do the right thing.

If you are committed to a renewable energy future, return on investment may be irrelevant. Creating a sustainable future, helping reverse costly global climate change, protecting the world for your children and your grandchildren, protecting the millions of other species that share this planet with us, setting a good example, leaving a legacy, helping the fledgling wind energy industry take off, and doing the right thing are all perfectly acceptable reasons for installing a wind system.

A Primer on
Wind Generators

One of the most important decisions you'll make when installing a wind energy system is the type of wind generator. Although there are a lot of companies manufacturing them worldwide, you'll find that your choice of models is limited—very few wind turbines are imported into the United States, for example. This chapter will help you understand your options and help you make a wise choice. I'll begin with the basics by examining the types of wind turbines on the market today and then examine the components of the most common turbines: horizontal-axis turbines. I'll also examine vertical-axis wind turbines and dispel the common myths about them. To help you sort through your options when shopping for a turbine, I'll discuss important features you should consider—features that could ensure many years of low-maintenance service. I follow this discussion with a brief exploration of homemade wind turbines.

The Anatomy of a Wind Turbine

Small wind generators come in many shapes and sizes, and they can be grouped into several different categories. You'll first need to select a turbine that is designed for the type of system you want to install—either a batteryless grid-tied system or a battery-based system. (The latter includes off-grid as well as grid-tied with battery backup.) Accordingly, you'll encounter two basic varieties: (1) turbines designed to charge batteries, appropriately called *battery-charging turbines*, and (2) turbines designed to connect to the grid. They are referred to as *batteryless grid-tied turbines*.

There's another distinction you will soon discover. It has to do with the orientation of the blades and shaft (if there is one) of the turbine. As just noted, most wind turbines are as horizontal-axis wind turbines, or HAWTs, briefly described

FIGURE 5.1. Upwind Turbine. A wind turbine like this Luminous Whisper 200 consists of one or more blades—usually three—attached to a central hub. When the wind blows, the rotor is located upwind from the tower. This is the most common type of small wind turbine. Credit: Dan Chiras.

in Chapter 3. Horizontal wind turbines consist of a set of blades attached to a central hub. The hub is connected to an alternator, either directly or through a shaft or gearbox. In either case, the axis of the rotor is oriented horizontally, hence the name.

The second, and less common type, are the egg beaters. These are known as vertical-axis wind turbines because the "blades" and shaft are oriented vertically. As you will soon learn, this type of turbine has been around for ages, but just doesn't perform as well as the more common, industry-wide standard HAWT.

As noted earlier, in this book, I focus most of my attention on horizontal-axis wind turbines. In this group, you will encounter two types of turbine: upwind and downwind. Upwind design is the most common. Figure 5.1 shows an example. You've probably seen many of them in your life. In these turbines, the rotor is located on the windward (upwind) side of the tower when the turbine is operating. These turbines all have tails that keep the turbine pointed into the wind. Downwind wind turbines are "tailless" turbines like the one shown in Figure 5.2. When the wind's blowing, the blades are downwind from the tower.

Horizontal-axis Wind Turbines

Horizontal-axis wind turbines vary somewhat in design but all upwind turbines consist of three main parts: (1) a rotor, (2) an alternator, and (3) a tail (Figure 5.3). Many modern turbines are equipped with three blades attached to a hub, covered by a nose cone. Together, the hub and blades form the rotor of the turbine (not to be confused with the rotor in the alternator). This entire assembly rotates when wind blows past the blades, hence the name "rotor." In older-style small wind turbines the rotor is attached to a shaft that's attached to an alternator, a device that produces AC electricity.

FIGURE 5.2. Downwind Turbine. Less common, downwind turbines have no tails. They typically consists of three blades attached to the alternator. When the wind blows, the rotor is located downwind from the tower. This turbine was once manufactured in Scotland by a company called Proven. Two models, a 3-kW and a 6-kW turbine machine, from the Proven line are now being manufactured by Kingspan in the UK. Credit: Dan Chiras.

Alternator vs. Generator

A generator is a machine that converts mechanical energy into electrical energy, either AC or DC. A generator that produces AC is called an AC generator or, more commonly, an alternator. Alternators in automobiles produce AC electricity which is then converted into DC electricity by a device known as a *rectifier*. Most, if not all, modern wind turbines contain alternators that produce variable AC electricity. However there are a few wind turbines that contain alternators equipped with rectifiers, just like the alternator in your car. They convert the AC to DC, which is then sent down the tower.

Alternators consist of two main parts, a set of stationary windings, known as the stator, and a set of rotating magnets, known as the rotor (Figure 5.4). In this type of turbine, when wind causes the blades of a turbine to spin, the shaft spins. The spinning shaft, in turn, causes the rotor of the alternator to spin, creating electricity. Wind turbines, therefore, first convert the kinetic energy of the wind into mechanical energy (rotation). The mechanical energy is then converted into electrical energy in the alternator.

More modern designs have done away with the shaft. The rotor of the turbine is attached directly to the rotor of the alternator, as shown in Figure 5.4. The blades of the turbine attach directly to a cylindrical metal "can" containing the magnets, the rotor of the alternator. These magnets spin around a set of stationary coils of copper wire, known as the windings. The windings constitute the stator of the alternator. The rotation of the magnets around the windings produces alternating current electricity in the windings via electromagnetic induction.

Alternators in many small wind turbines contain a special type of magnet, known for producing very strong magnetic fields. These are known rare earth magnets. Rare earth magnets contain the rare earth element, neodymium, and two more common elements, iron and boron. They produce a much stronger magnetic field than conventional iron (ferrite) or ceramic magnets used in older small wind turbines. The stronger the magnetic field, the greater the output of a wind generator, all things being equal.

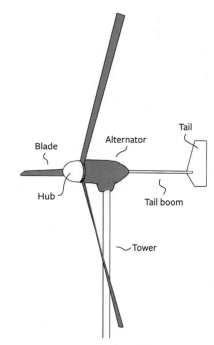

FIGURE 5.3. Anatomy of a Wind Generator. A wind turbine consists of blades attached to a hub. They form the rotor. The rotor in this drawing is attached via a shaft to the rotor of the alternator. When it spins, it produces electricity. Credit: Anil Rao.

FIGURE 5.4. Anatomy of an Alternator. This simplified diagram of alternator consists of a stationary set of windings, the stator, in which electricity is generated. The rotor of this alternator contains magnets that spin around the windings of the stator when the wind blows. These are known as permanent magnet alternators. Some wind turbines like large, commercial turbines contain electromagnets. (Electromagnets are so named because they are created by running DC electricity through coils of copper wire. This creates a magnetic field around the coils.) Credit: Anil Rao.

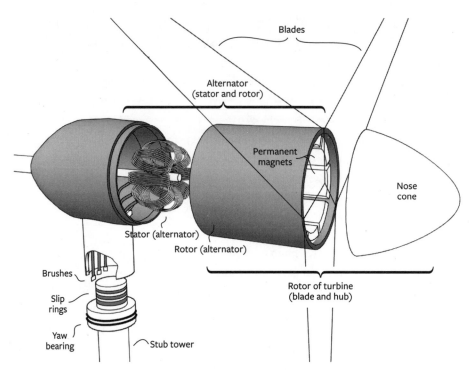

Generators that use iron, ceramic, and rare earth magnets, as opposed to electromagnets, are referred to as *permanent magnetic generators*. The magnets are mounted on the rotor of the generator (the part that spins).

Electricity leaves the alternator via wires that attach to the stator. In most turbines, like the one shown in Figure 5.4, wires from the alternator terminate on metal brushes. The brushes, in turn, contact slip rings, brass rings located near the yaw bearing. The yaw bearing is a component that allows the turbine to turn in response to changes in the wind direction. The brushes transfer electricity from the alternator to the slip rings. The slip rings, in turn, connect to wires that run down the length of the tower. Electricity therefore flows from the alternator to the brushes to the slip rings and then down the tower to the controller.

Wind generators such as the ones shown in Figures 5.3 and 5.4 are known as direct drive turbines. That's because the rotor (blades and hub) are attached directly to the rotor of the alternator. As a consequence, the rotor and the alternator turn at the same speed. Although virtually all modern residential wind turbines are direct drive, large commercial wind turbines and a few of the small wind turbines contain gearboxes installed between the rotor of the turbine (blades and hub) and the alternator. These are known as indirect drive or gear-driven turbines. (For more on this, see accompanying sidebar.)

Another important component of upwind horizontal-axis wind turbines is the tail. The tail typically consists of a boom and a vane. The tail boom connects the tail vane to the body of the turbine.

Tails are passive directional control devices. They keep the rotor of the wind turbine pointing into the wind. If the wind direction shifts, the tale vane turns the turbine into the wind, ensuring maximum electrical energy production. In the language of the wind industry, the movement of a wind turbine on a tower as it tracks the wind direction is referred to as *yawing*.

Although upwind turbines dominate the market, several manufacturers produce downwind turbines, among them Xzeres (Skystream) and Ventera.

In downwind turbines, the blades are oriented downwind from the tower; this occurs because the rotor is located several feet in front of and slightly to the side of the yaw bearing of the turbine (located where the turbine attaches to the tower). When the wind blows, it pushes the rotor and the generator to the downwind position. Yawing may be controlled by the blades of some downwind turbines. The blades of the Proven wind turbines (no longer available), for instance, are hinged. When the wind blows and the blades spin, the hinged blades bend very slightly, forming a shallow cone. Coning in these turbines helps orient the rotor downwind.

Downwind turbines work well. However, if the wind dies down and then reverses direction, they can get stuck in the upwind position (with blades upwind from the tower). When oriented upwind from the tower, downwind turbines will not spin and will not generate electricity. I have noticed this many times in my Skystream 3.7. It's very frustrating to watch. I have tugged on the guy cables to jiggle the turbine, hoping that the turbine might reorient itself into the wind, but to no avail. I then go inside and sometime later find the turbine oriented correctly and working fine.

Getting stuck upwind is not unique to the Skystream; it's characteristic of all downwind passive yaw small turbines. However, most experts agree that this isn't a huge problem. It is more of an idiosyncrasy—a mild annoyance to the owner. As the wind speed increases or shifts direction, the turbine will align itself properly.

As noted earlier, there's also another type of wind turbine that is getting a lot of attention. It is known as a *vertical-axis* wind turbine. To learn more, check out the accompanying textbox. The title tells it all.

Direct Drive vs. Gear Drive

Large commercial generators and several small wind turbines are equipped with gearboxes interposed between the rotor and the generator. They increase the speed at which the alternator spins, thus increasing the output of the alternator. This, in turn, allows the alternator to be much smaller and also reduces rotor speed (blade speed) making the turbines quieter. In the Jacobs 31-20 (a 20 kilowatt-hour-rated turbine with a 31-foot-diameter rotor), the gear ratio is 6:1. That means that the alternator turns six times for every rotation of the rotor (the blades). Gearboxes are also found in the rebuilt Vestas V-15 and V-17 turbines.

The Dubious Promises of Vertical-axis Wind Turbines

As Mick Sagrillo writes in his column on small turbines in the American Wind Energy Association's newsletter *Windletter*, "Yet another announcement about a technological breakthrough in the wind turbine industry has just arrived in my e-mail inbox. Invariably the device being unveiled is a vertical-axis wind turbine. Included with the excited chatter is an article or two published by some local newspaper," raising interest—and, I say, false hope.

As shown in Figure 5.5, the blades of a vertical-axis wind turbine (VAWT) are attached to a central vertical shaft, hence their name. When the blades of a VAWT spin, the shaft spins. The shaft is attached to an alternator generally located at the bottom of the shaft, sometimes even at ground level.

"Vertical-axis wind energy devices have been around for a long time…about 3,000 years," notes Mick. These designs are so simple that many tinkerers/inventors develop unique designs and prototypes. The internet is flooded with examples.

Proponents of VAWTs tout a number of supposed "advantages" over HAWTs that may, to the uninformed, seem to give this technology a decisive edge. One supposed advantage is that VAWTs can capture wind from any direction. The turbines don't need to be oriented into the wind, as do HAWTs . One benefit of this, say proponents, is that VAWTs are immune to the turbulence that wreaks havoc with HAWTs.

Another supposed advantage, say proponents, is that VAWTs don't need tall towers. They can be mounted close to the ground—even on top of buildings—where they capture ground-level winds (Figure 5.6). This, say supporters, eliminates the need for tall and costly towers and the need to obtain the zoning variances sometimes required to install horizontal-axis wind turbines on tall towers.

Another supposed advantage of the VAWTs stems from the fact that the generator can be mounted at ground level. This, say proponents, makes it easier to

FIGURE 5.5. Vertical-axis Wind Turbine. Although there's a lot of interest these days in vertical-axis wind turbines, they are mounted at ground level, which exposes them to unproductive low-speed winds. As a result, they don't produce much energy. And that's just one of their problems! Credit: Sandia National Laboratories.

access and repair the generator should the need arise. There's no need to climb the tower—or lower a wind generator to the ground—to perform routine maintenance or to replace damaged or worn parts. These activities take time and can be risky to inexperienced individuals.

While these arguments cast VAWTs in a good light and are intended to convince us of their merit, they are either invalid or greatly exaggerated. Years of experience with VAWTS has been discouraging, to say the least. "Hundreds of commercial VAWTs were installed in California in the late 1980s and early 1990s," consultant Bob Aram reminds us. "They all failed and were removed from service. These were not experimental units, but production units."

Proponents argue that VAWTs are better than HAWTs because they can capture wind from any direction. That's a bit of exaggeration because HAWTs *also* capture wind from any direction; they shift orientation as the wind direction changes. During directional shifts, they may lose a little power, but the losses are pretty small. So, while the VAWT does have an advantage in dealing with shifts in wind direction, it is not much.

The second supposed advantage of VAWTs— that they don't need tall towers—is true, but false reasoning. Although VAWTs can capture ground-level winds (just like any turbine installed on a too-short tower), ground-level winds are subject to friction (causing ground drag) and generally experience a lot of turbulence. Both ground drag and turbulence

FIGURE 5.6. Although these rooftop wind turbines in St. Louis, Missouri, look cool, they produce very little electricity. They can also produce a substantial amount of vibration. In essence, they become costly, annoying roof ornaments. Credit: Dan Chiras.

in the lower-level winds where VAWTs are mounted diminish the power available to a turbine—so much so, that there is very little extractable energy in them. The lower the wind speed, the less electricity a turbine will produce. Just because a VAWT can be mounted at ground level doesn't mean it's a good idea. It won't produce much electricity.

VAWTs have other inherent problems that proponents fail to mention. For one, the VAWTs are less efficient than horizontal-axis wind turbines. "For a given swept area," NREL's Jim Green notes, "they just don't extract as much wind energy as a well-designed HAWT."

Another problem is that the blades of VAWTs are prone to metal fatigue. This is created by centrifugal forces as the blades spin around the central axis. The vertically oriented blades used in some early models, for instance, twisted and bent as they rotated in the wind. This caused the blades to flex and crack—that is, to fatigue, in industry lingo. Over time, repeated flexing caused the blades to crack and eventually break apart, sometimes leading to catastrophic failure. Because of this, VAWTs have proven less reliable than HAWTs, and the early VAWTs have long since been abandoned.

Many VAWTs also require large bearings at the top of the tower to permit rotation of the shaft. When the top bearings or the blades need replacement, you've got a job on your hands. Another problem with VAWTs is that many models require very thick and costly steel cables to support them. This contributes to the higher cost of electricity from VAWTs compared to HAWTs.

While many modern VAWT inventors show videos of their turbines spinning, which is typically enough to convince unwitting reporters and the easily duped and effervescently enthused general public of their viability, it's not spinning blades that matter. What matters is energy output—how many kilowatt-hours a wind turbine generates when mounted on its tower. Because average wind speeds are very low at ground level at most sites and because VAWTs are less efficient than HAWTs, VAWTs won't produce much electricity. Don't forget, too, that eddies can form behind buildings and trees at ground level, resulting in dead air spaces, regions where the average wind speed is close to zero. Place a VAWT in a location such as this and you've just installed an expensive lawn ornament.

VAWTs are less reliable and less efficient than HAWTs, and "these deficiencies are only made worse by mounting at ground level or on the top of buildings," notes Jim. What electricity they produce turns out to be more expensive. All in all, they just don't stack up against horizontal-axis wind turbines. The few advantages they offer cannot counter their serious disadvantages. If you are thinking about buying one, forget it. If you are thinking about inventing one, save your time.

Improvements in Small Wind Turbines

Although wind turbines in the 1970s and early 1980s had problems that gave the wind industry a black eye, market forces have eliminated many poor designs and marginal producers. Moreover, small wind turbines have improved dramatically. In fact, most small horizontal-axis wind turbines on the market today have been redesigned for simplicity, durability, and high performance. Manufacturers have reduced or eliminated moving parts and parts that are subject to wear, such as brushes in alternators. (Brushes are still used in the yaw bearings to transfer electricity from the alternator to the wires leading down the tower.) They've substituted materials like high-strength rare earth magnets for ordinary ferrous magnets. And they've spent a considerable amount of time researching and testing blade materials and design to improve their efficiency. The sidebar below summarizes many of the improvements in turbines.

Costly research and development, often carried out in conjunction with the US Department of Energy's Wind Energy Technology Center in Golden, Colorado, (part of DOE's National Renewable Energy Laboratory) has led to the production of much more durable wind turbines, many of which can provide years of relatively trouble-free service. In fact, if properly sited, installed, and maintained, a high-quality wind turbine could last 20 to 30 years. That's no small feat, given the fact that a wind turbine in a good site operates as many hours in a year as an automobile operates over its lifetime. Improvements in the design and construction of small wind turbines have also resulted in longer warranties, typically five years. Bergey offers a couple of wind turbines with unprecedented 10-year warranties. How many products do you know that come with 10-year warranties?

Trends in Wind Turbine and Wind System Design and Production

- Improved blade design and materials leading to greater strength and higher performance
- Greater generator efficiency, in large part by incorporating rare earth permanent magnets
- Reintroduction of induction generators that produce grid-compatible electricity without the use of an inverter
- Design for lower average wind speeds to improve output
- New and improved methods of overspeed control
- Reductions in rotor speed to reduce sound from blades
- Improvements in the efficiency and reliability of grid-connected inverters

Modified from Jim Green, Wind Energy Technology Center, National Renewable Energy Laboratory, Golden, Colorado

Although improvements have resulted in much better wind turbines, buyers should not assume that all wind generators on the market today are the same. As you will see shortly, a number of less expensive light- and medium-duty turbines were not designed and built to last. The predominant design principle seemed to be affordability. While cost is important, turbines designed principally with up-front cost in mind typically don't last long, especially if installed in high or turbulent wind sites or sites that experience a lot of storms with strong winds.

Over the years, there have been some cheap imports on the market that have not lived up to the manufacturers' expectations. Many of them failed to operate at all. Beware of new entries into the field.

The Main Components of Wind Turbines

To help you select a wind turbine that's built to last, let's take a closer look at the main components of modern wind turbines, starting with blades and generators.

Blades

Blades are a vital component of a wind turbine. Manufacturers make blades from several different types of materials.

Materials

Although wood was once commonly used to make blades for small wind turbines, it has been replaced by plastics and composites such as fiberglass. In fact, we're not aware of any companies that currently manufacture blades exclusively from wood. It's just too expensive and time-consuming.

To ensure years of reliable service, the blades of most modern wind turbines are made of extremely durable synthetic materials, various types of plastic or composites—plastic reinforced with fiberglass or carbon fibers, for example. These synthetic blades typically last between 10 and 20 years. Table 5.1 lists some examples.

Table 5.1. Composition of Small Wind Turbine Blades

Company	Blade Material
Abundant Renewable Energy	Fiberglass
	Fiberglass
Bergey Windpower	Fiberglass
Wind Turbine Industries	Polypropylene (WT600)
Southwest Windpower	Polypropylene reinforced with fiberglass, Fiberglass

Why use plastic?

Manufacturers have found that fiberglass and other plastics are cheaper than wood, in part, because they require a lot less labor to make. Unlike the metal blades once used in small wind turbines, plastics blades don't interfere with television, satellite TV, or wireless internet signals.

Two Blades or Three?

Most modern residential wind turbines—both upwind and downwind models—have three blades. Although a few manufacturers produce two- and six-blade turbines (for sailboats), three-blade wind turbines are the industry standard.

There's a reason for this—the number of blades used in a wind turbine is a compromise between efficiency, sound, longevity, and cost. Let's begin with efficiency.

Interestingly, the most efficient number of blades is one. Contrary to what you might think, single-blade wind turbines are able to convert wind energy into electrical energy more efficiently than any other configuration. Unfortunately, single-blade designs are difficult to balance when operating. As a result, you won't see single-blade turbines being offered by wind turbine manufacturers.

Two- and three-blade wind turbines operate at about the same efficiency (two-blade turbines are slightly more efficient than three-blade models). That's because, all other factors being equal, two-blade turbines spin faster than three-blade turbines and produce slightly more electricity. Two-blade turbines cost less to build than three-blade turbines, because they require less material and less labor. Because they spin faster, a manufacturer can use a smaller, less expensive generator. All this adds up to a less expensive turbine. If all this is true, then, why don't two-blade wind turbines dominate the market?

One reason is that three-blade wind turbines are quieter. Why?

One reason three-blade turbines produce less sound is that the more blades a turbine has, the lower the rotor speed. As a general rule, the lower the rotor speed, the quieter the turbine. (It should be pointed out that other factors like overspeed protection, discussed shortly, come into play when sound levels are concerned, but most of the sound generated by a small wind turbine is from the blades.)

Another reason three-blade turbines are quieter is that they don't produce yaw chatter, a phenomenon that plagues two-blade turbines as they turn in the wind. Yaw chatter is a sound created by the vibration of the blades in a two-blade turbine as the turbine is yawing (turning with the wind). Chatter also increases wear and tear on a turbine, which can lead to more maintenance and repair. It can also considerably shorten the life of a turbine.

Three-blade rotors eliminate chatter for reasons beyond the scope of this book. Although they cost a bit more to manufacture, the quieter, three-blade turbines are more popular among customers. People also like them because they suffer less wear and tear and therefore last longer. Although initial cost may concern some readers, I strongly urge prospective wind system owners to focus on quality. Longer-lasting wind turbines may cost more upfront, but your return on investment can be significantly higher.

Generators

Most residential wind turbines on the market contain permanent magnet alternators, described earlier, although larger turbines in the small turbine range, like the remanufactured Jacobs 31-20, typically incorporate electromagnets. As noted in Chapter 3, most wind turbines produce variable AC, initially. It is converted (rectified) to DC electricity by a device known as a *rectifier*. The rectifier may be located in the wind turbine itself or in the controller. Controllers are typically mounted next to the inverter, which is located in or near the home or business. DC electricity is then sent to the battery bank in battery-based systems or to the inverter in grid-tied systems.

As you may recall from Chapter 3, the charge controller in battery-based systems not only converts AC to DC, it regulates the flow of electricity to the batteries, preventing overcharge. If the batteries of an off-grid system are full, surplus electricity is fed into a diversion load, a resistance heater located in a water heater or a space heater. In grid-connected systems with battery backup, the surplus electricity is backfed onto the grid.

In off-grid systems or utility-connected systems with batteries, batteries deliver DC electricity to the inverter. The inverter converts DC to AC and boosts the voltage from 12, 24, or 48 volts, the voltage of the turbine and battery bank, to 120 or 240 volts, the voltages required in most households and businesses. In utility-connected systems, the inverter converts the DC back to AC but to a much "tamer" version, one that matches grid power—60 cycle per second 240-volt electricity in North America (in synch with the utility's grid electricity). The electricity is then delivered to active loads in the house via the main service panel. The excess is backfed onto the electric grid.

Also noted briefly in Chapter 3, some manufacturers are now producing small turbines with induction generators. These wind turbines produce grid-compatible AC power without the use of an inverter. An induction generator looks a lot like the electric induction motors widely used in modern society. When spun faster than its normal operating speed by the blades of a wind turbine, an induction generator produces AC electricity. Because an induction generator synchronizes with the grid—that is, produces electricity at the same frequency and voltage—no inverter is required. (However, other electronic controls are needed.)

With these basics in mind, we now turn our attention to specific features to look for when buying a wind turbine.

What to Look for When Buying a Wind Turbine

If wind turns out to be a viable option for you, you'll need to select a wind turbine. While there are many turbines on the market, your electrical demand will narrow the field considerably. That is, once you have determined your annual energy demand and the average wind speed on your site, you can select a wind turbine that will produce enough electricity to meet your demands. You learned how to do that in Chapter 4. Table 4.5, you may recall, showed your options. Once you narrow your search to one or two turbines, how do you choose the right one?

Manufacturers provide a plethora of technical data on their wind turbines that can be used to make comparisons. Unfortunately, most of it is useless. Further complicating matters, "There can be a big difference in reliability, ruggedness, and life expectancy from one brand to the next," according to Mike Bergey.

So how do you go about selecting a reliable wind turbine?

Although wind turbines can be compared by many criteria, there are only a handful that really matter: (1) swept area, (2) annual energy output, (3) durability, (4) cost, (5) certification, (6) governing system, (7) shut-down mechanism, and (8) sound. Let's take a look at each one.

Swept Area

In wind energy, blade size is a critical feature. To get the most out of a wind turbine— to produce the most electricity at the lowest cost—select a wind turbine with long blades. As discussed in Chapter 4, the longer the blade, the greater the swept area.

Swept area is the area of the circle described by the spinning blades of a turbine. Because the blades collect energy from the wind, the swept area is the collector area of a turbine. It is akin to the square footage of a solar-electric array. Doubling the area of a solar-electric array (the area exposed to the Sun) doubles its output. The same holds true for wind turbines. As a rule of thumb, then, the bigger the swept area, the more energy you'll be able to capture from the wind. Swept area allows for easy comparison of different models. When comparing two or more turbines, you will find in many cases the greater the swept area, the greater the energy output.

Look for the models that have the greatest swept area. As Hugh Piggott of Scoraig Wind Electric in Scotland puts it, "Swept area is easier to measure and harder to lie about than performance."

Swept area is determined by rotor diameter. The rotor diameter is the distance from one side of the circle created by the spinning blades to a point on the opposite side, or about twice the length of the blades.

When comparing wind turbines, the blade length and, consequently, rotor diameter are a pretty good indicator of how much electricity a turbine will generate. Although other features such as the efficiency of the generator and the design of the blades influence energy production, for most turbines they pale in comparison to the influence of rotor diameter and swept area. If you are comparing wind turbines, and one has a 3-foot (1-meter) blade and the other has a 6-foot (2-meter) blade, the latter will generally be the best choice.

Manufacturers list the rotor diameter in feet or meters—often both. Never forget: the greater the blade length, the greater the rotor diameter, and the greater the swept area—that is, the greater the amount of wind a rotor can intercept and convert to electricity.

Fortunately, most manufacturers also list the swept area of the rotor. Swept area is presented in square feet or square meters—sometimes both. Table 5.2 lists the swept area of numerous residential-sized wind turbines. Table 5.3 lists this information for much larger wind turbines that still fit within the grouping "Small Wind Turbines." Table 5.4 shows their estimated annual energy output at various wind speeds.

To appreciate the importance of rotor diameter and hence swept area, take a look at Figure 5.7. It illustrates the theoretical increase in power production in a 22.4 mile-per-hour (10 meter per second) wind. Power production theoretically increases from 0.4 kilowatt-hours to 1.6 kilowatt-hours with a doubling of rotor diameter (from 1 meter to 2 meters). When rotor diameter increases from 2 to 4 meters, theoretical power output increases from 1.6 kilowatt-hours to 6.4 kilowatt-hours. (Remember that these are theoretical estimates. Other factors such as turbine and blade efficiency and air density influence electrical output.)

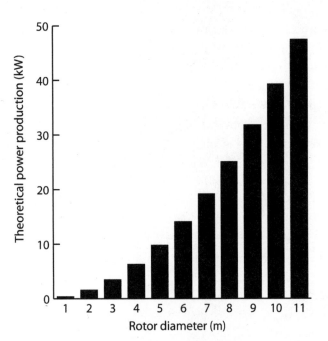

FIGURE 5.7. Electrical Production vs. Rotor Diameter. Theoretical power production of a small wind turbine at a wind speed of 22.4 miles per hour (10 meters per second). Notice the effect that increasing rotor diameter has on power production. Credit: Anil Rao.

Table 5.2. Features of Selected Small Wind Turbines in the 1 to 10 kW Range

Turbine	Rated Power at 11 m/s (kW)	AEO at 5 m/s (kWh)	RPM at Rated Power	Rotor Swept Area (ft²)	Top of Tower Weight (lbs)	Warranty	Years in Production as of June 2016
Bergey Excel 1	1.00	1,110	490	53	75	5	16
Luminous Whisper 200	0.99	2,052	1,200	64	66	2	7
Pika T 701	1.5	2,420	400	76	93	5	2
Xzeres Skystream 3.7	2.10	3,420	330	117	170	5	3
Luminous Whisper 500	3.10	5,568	800	177	154	2	7
Luminous Windistar 4500	4.31	7,800	450	177	249	2	3
Bergey Excel 6	5.50	9,920	400	325	772	5	4
Ventera VT-10	9.30	12,772	280	380	580	5	8.5
Bergey Excel 10	8.90	13,800	400	414	1,200	10	33
Bergey Excel R	7.50	13,800	400	414	1,200	10	33
Xzeres 442SR	9.17	15,329	150	442	1,600	10 turbine 5 controller	5

Table 5.3. Features of Selected Larger Small Wind Turbines According to Manufacturers

Turbine	Rated Power at 11 m/s (kW)	AEO at 5 m/s (kWh)	RPM at Rated Power	Rotor Swept Area (ft²)	Top of Tower Weight (lbs)	Warranty	Years in Production as of June 2016
WTI Jacobs 31-20	12	16,630	175–185	755	2,000	5	31
Osiris 10	10	23,704	120	797	1,870	5	7
Ecocycle 25	23	37,229	90	1,347	4,960	2 with 5 extended	6
Endurance Wind Power E-3120	56.8	116,935	42	3,120	8,800	5	7
Northern Power NPS-100C-24	90.2	196,000	50	4,867	15,300	2	17

Table 5.4. Estimated Annual Energy Output in kWh of Selected Larger Small Wind Turbines According to Manufacturers

Turbine	8 mph	9 mph	10 mph	11 mph	12 mph	13 mph	14 mph
WTI Jacobs 31-20	5,100–7,800	7,450–11,375	10,420–15,900	13,900–21,225	17,742–27,075	21,950–33,470	26,250–38,950
Osiris 10	8,250	13,929	18,824	22,740	27,530	31,340	35,734
Ecocycle 25	11,230	18,992	25,625	34,160	44,471	53,775	64,700
Endurance Wind Power E-3120	41,516	65,214	88,913	112,611	137,334	162,289	186,458
Northern Power NPS-100C-24	80,000	113,000	150,000	189,000	228,000	267,000	305,000

Annual Energy Output

Although greater swept area generally translates into greater electrical production, another, even more useful, measure is the annual energy output (AEO) at a newly adopted industry-standard rated wind speed—11 meters per second (24.61 miles per hour). AEO is reported in kilowatt-hours per year at 11 m/s, as you can see in Tables 5.2 and 5.3. Throughout most of small wind industry's history, manufacturers rated their wind turbines at different speeds, making it difficult to compare one to the other. The new standard allows us to compare one turbine to another.

Although the standard 11 m/s is useful for comparing turbines, it is still useful to know a turbine's expected AEO at various wind speeds so you can select one that will produce the electrical energy you need on the top of the tower you selected. Tables 4.5 and 5.3 list AEOs for numerous small wind turbines at a number of speeds.

AEO is determined by the manufacturers. Some are conservative in their calculations so as not to disappoint owners with exaggerated estimates of energy production. Other manufacturers are more "optimistic" in their energy estimates, but their products may fail to deliver.

Despite this caveat, AEO provides a useful means of comparison. As shown in Table 4.5, the Ventera VT-10 should produce approximately 5,037 kWh at sites with an average wind speed of 8 miles per hour, and 11,625 kWh at sites with an average wind speed of 11 miles per hour (this is according to the manufacturer's estimates, of course). As shown in Table 5.4, the Osiris 10 should produce approximately 8,250 kWh at sites with an average wind speed of 8 miles per hour and 22,740 kilowatt-hours per year at sites with an average wind speed of 11 miles per hour, again according to the manufacturer's estimates.

Durability: Maximum Design Wind Speed and Weight

Another extremely important criterion to consider when comparing wind turbines is durability. One measure of durability you'll see in advertisements is the maximum design wind speed. Maximum design wind speed is the wind speed a turbine can supposedly withstand without sustaining damage. At first glance, you'd think that this might be a good parameter by which to compare turbines. It stands to reason that a wind turbine designed to withstand 130 mile-per-hour winds ought to be more rugged than one designed to withstand 110 mile-per-hour winds.

The problem with maximum design wind speed is that a wind turbine designed to survive a wind speed of, say, 120 mph is either not tested at that speed (or not repeatedly tested at that speed) by manufacturers. Another problem with using maximum design wind speed as a guiding parameter is that wind generators are more often damaged by turbulence than by high-speed winds.

A far better measure of durability is top-of-tower weight—how much a wind turbine weighs. See Tables 5.2 and 5.3 for a list of turbines and their top-of-tower weight. As a rule, heavyweight wind turbines tend to survive the longest—sometimes many years longer than medium- or lightweight turbines. So, when comparing two otherwise comparable turbines, the heavier wind turbine is probably more durable.

Cost

Although swept area, annual energy output (AEO), and top-of-tower weight are the primary criteria one should consider when shopping for a wind turbine, cost is also important. For many potential small wind energy producers, initial cost is often used to choose a wind turbine. Be forewarned, however: quality is usually reflected in the price. Generally, the heavier and more durable a wind turbine is, the more you'll pay. Remember, however, that you get what you pay for. Producing electricity on a precarious perch 80 to 165 feet above the ground isn't a job you want to relegate to the lowest bidder, which is invariably the one with the lightest turbine.

Don't forget that there's more to a wind system than the turbine. You'll need to purchase a tall tower and pay for installation, unless, of course, you install the tower yourself. Even then, you'll need to pay for concrete, rebar, and equipment to excavate the foundation and anchors. You'll also need to run electrical wire from the turbine to the house and purchase an inverter (although they're included in most batteryless grid-tied wind turbines). If you're going off-grid or want battery backup for your grid-connected systems, you'll also need to buy batteries. All of this will add to the cost. The cost of the turbine may range from 10 to 40 percent of the total system cost.

Certification

When buying a turbine is a good idea to check to see if the turbine has been certified by the Small Wind Certification Council (SWCC). The SWCC is a relatively new organization formed to provide independent third-party testing of wind turbines that will assist buyers in evaluating wind turbines. "Certification can help prevent unethical marketing and false claims, thereby ensuring consumer protection and industry credibility," according to the SWCC.

The SWCC certifies small and medium wind turbines using a set of criteria established by a group of wind energy experts, including manufacturers. For small wind turbines, SWCC Certification is based on an evaluation of data it collects during field tests—measurements from wind turbines mounted on towers at test sites. The evaluation results in an estimate of AEO (annual energy output)—total kilowatt-hours of electricity produced—at a site with an average wind speed of 5 m/sec or 11.2 mph. It also includes a measure of sound levels 60 meters from the tower when the turbine is operating at 5 m/s. SWCC also provides an industry standard for rating wind turbines: the power output (in watts) at 11 m/s. Moreover, SWCC determines the peak power—the maximum output of a wind turbine. (You can go online to view the data on a handful of small and medium wind turbines certified to date.)

It is worth noting that data on wind turbines that have not been certified are provided to buyers from the manufacturer. They may be derived from field tests or wind tunnel tests at conditions that make their turbines look better than others. So, be wary of data provided from manufacturers. Be certain you know how it was derived.

Besides performance, turbines are tested for safety and durability. This data is included in a summary report that is available to the public via the internet once a turbine has been certified.

All tests are performed by accredited and non-accredited testing labs. According to SWCC, "Test reports from accredited testing organizations such as the National Renewable Energy Laboratory (NREL) will require the minimum level of scrutiny from the SWCC. Testing performed by non-accredited organizations will require a higher level of scrutiny to independently verify the test setup complies with the Standard, the competence of the organization, and the quality of the test reports."

Governing System

A vital component of all wind generators is the governing system, or overspeed control. These are systems designed to prevent a wind generator from spinning too fast once wind speeds exceed a safe level. This feature prevents wind turbines from burning out or breaking apart in high winds. All overspeed controls govern by slowing down the rotor when the wind reaches a certain speed, known as the governing wind speed. Why is this necessary?

As wind speed increases, the rotor of a wind turbine spins more rapidly. The increase in the revolutions per minute (rpm) increases electrical output. Although electrical output is a desirable goal, if it exceeds the turbine's rated output, the generator could overheat and burn out. In addition, centrifugal forces in high wind speeds exert incredible forces on wind turbines that can tear them apart if the rpm is not limited.

Designing a wind turbine to handle all that force would make an extremely heavy machine that would likely not perform well in moderate winds found in most locations, the winds we capture to produce most of our electricity.

A governing system allows the turbine to survive high winds, so it is available for the next capturable wind. Not all wind turbines come with governing mechanisms, however. Many of the smallest wind turbines, the microturbines, with rated outputs of around 400 watts, for example, have no governing mechanisms (Figure 5.8). (These turbines are too small to produce an amount of electricity significant enough for applications covered in this book.) Larger wind turbines rated at 1,000 or higher come with some form of overspeed control. Two types are commonly found: furling and blade pitch.

FIGURE 5.8. Microturbines. Many microturbines like the Marlec (shown here) have no governing mechanism to slow the rotor in high winds. They rely on the relatively low rotor speed and rugged construction to endure high winds. Source: Marlec.

Furling

Most manufacturers protect their wind turbines by furling. Furling is an ancient term that originated in the days when windmills were built with fabric-coated blades. The miller who operated the windmills climbed onto the blades and rolled up, or furled, the fabric so they would no longer turn. This protected the windmills from damage in high winds.

Today, the term is used to describe a very different process that achieves the same result. Furling is accomplished in one of two ways, both of which shift the position of the rotor (hub and blades) relative to the wind, effectively turning the blades out of the wind. This, in turn, decreases the amount of rotor swept area (known as the frontal area) that intercepts the wind. Reducing the swept area reduces the speed at which the rotor turns and the energy is collected. Slowing the rotor protects the wind turbine from damage.

Manufacturers employ two main types of furling: horizontal and vertical. In horizontal furling, the rotor turns out of the wind by turning sideways. For this reason, horizontal furling is also known as *side furling*. In vertical furling, the rotor rotates upward with the same effect. *Angle furling* is a combination of the two.

FIGURE 5.9. Side Furling. This wind turbine is not broken, it is side furling in a high wind, to slow the rotor and protect the machine from damage. Credit: Mick Sagrillo.

FIGURE 5.10. Vertical Furling. This is a more technologically complicated form of furling than side furling, and therefore uncommon. It can also result in a lot of unwanted sound. I can find no traces of the company that once produced this small wind turbine. Credit: North-Wind Power.

Horizontal or side furling is achieved, in part, by hinging the tail. In side-furling turbines, a hinge is located between the tail boom and the body of the turbine. As you can see from Figure 5.9, the turbine is also slightly offset from the yaw axis—that is, the yaw bearing is attached to the side of the turbine body, not its center so the turbine is not directly over the tower.

Because the turbine is offset from the yaw axis, the force of the wind on the blades tends to rotate the turbine around the yaw axis. However, the tail resists this rotation and keeps the rotor facing into the wind.

In light winds, the forces on the rotor and tail are small, and the wind holds the tail in its normal position—straight behind the turbine. However, in strong winds, the increasing forces on the rotor overcome the force of the wind on the tail. Since the tail creates more force than the offset rotor, the tail stays mostly aligned with the wind and the turbine turns away from the wind. As a result, the turbine literally folds on itself. As just noted, this reduces the frontal area of the rotor, the amount of swept area in contact with the wind, which slows the rotor. When the wind speed diminishes, however, the turbine unfolds, returning to its normal operating position.

Vertical furling is much less commonly used. It is achieved by moving the hinge in front of the yaw axis and rotating it slightly. In high winds, the force of the wind tilts the rotor up and to one side, while the tail stays oriented downwind. As in side furling, this reduces the frontal area of the rotor and reduces its speed. The lifting force of the wind is opposed by the weight of the rotor and generator, so no spring is needed. (Figure 5.10).[1]

When fully furled, the rotor of a vertical furling turbine resembles a helicopter rotor—and it can be

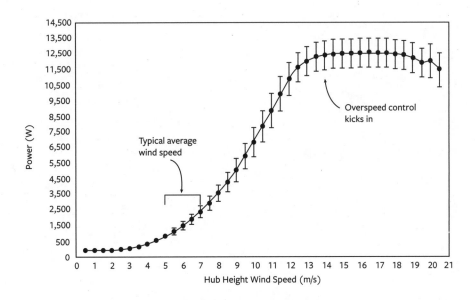

FIGURE 5.11. Power Curve. The electrical production (in watts) of the Bergey Excel 10 wind turbine. Like that of many other turbines, power output plateaus or declines in high wind speeds as a result of overspeed controls designed to protect wind turbines from damage. Data for this graph came from the Small Wind Certification Council's Summary Report. Credit: Forrest Chiras.

quite noisy, too. When wind speed declines, the rotor returns to its normal operating position. Shock absorbers are often used to ease the rotor back into position.

Furling reduces the amount of energy collected by the rotor. Although electrical output may continue, it typically occurs at a lower rate, as shown in the power curve in Figure 5.11. This graph shows the watts produced by a wind turbine at various wind speeds. Notice how power production drops off once the turbine furls.

Changing Blade Pitch

Another type of overspeed control involves a change in blade pitch—that is, a change in the angle of the blades—to reduce rotor speed. Blade pitch changes are used in propeller-driven airplanes to control power.[2] It works the same way in wind turbines. When wind speed increases to the upper limit of the operating range of the turbine, blade pitch changes automatically.

Changing the angle of the blade reduces its aerodynamic lift. Reducing lift, in turn, reduces rotor speed. The greater the wind speed above the operating range of the turbine, the more the blades rotate (pitch). The greater the change in pitch, the lower the lift, the lower the rotor speed.

Blade pitch control is found in a number of small wind turbines, among them Kestrel wind turbines, manufactured

A Word on Governors

Most manufacturers incorporate one type of governor, but some incorporate two mechanisms to achieve this end. For example, a turbine manufacturer may incorporate side furling *and* dynamic braking for overspeed control (dynamic braking is described shortly).

FIGURE 5.12. Pitch Control: Blade-Actuated Governor. Numerous ingenious methods of blade pitch control have been devised. In this turbine, a Jacobs 31-20, the springs are part of a complex and effective blade pitch control mechanism. Credit: Dan Chiras.

by a subsidiary of Eveready, Ltd., the South African company that bring us batteries. Blade pitch turbines are also manufactured by Jacobs. Pitch control typically requires springs, gears, and weights ingeniously arranged to produce the desired effect (Figure 5.12).

Blade pitch functions admirably, but is not as widely used as horizontal and vertical furling mechanisms. Why?

For many small wind turbine manufacturers, pitch control has proven too costly. Some even argue that pitch control mechanisms are less reliable and require more maintenance than the average homeowner is inclined to perform.

As in many issues, universal agreement on this subject is lacking. Many people who own and operate turbines with blade pitch control argue that their turbines operate better than furling turbines.

Mick Sagrillo sides with proponents of blade pitch governors. Even though there are more moving parts, he prefers blade pitch governors over furling anytime! Although furling mechanisms may be cheaper, furling is less reliable. "Furling may or may not work," he says. In addition, most furling mechanisms in use today don't slow the blades as they are supposed to and therefore don't provide the protection they're supposed to offer. Why have a governor if it's unreliable?

On the issue of maintenance, Mick does admit that blade pitch mechanisms need to be rebuilt every 10 to 15 years. However, furling governors require similar

maintenance in the same time frame. The bearings and bushings that are part of the passive furling mechanism must also be replaced every 10 to 15 years, although they are cheaper and easier to replace or rebuild than a pitch control system.

Both furling and blade pitch governors control rotor speed in high winds. However, blade pitch governors seem to give better control of speed and may be more reliable than furling. Bottom line: although furling mechanisms are cheaper, cheap is not necessarily better when it comes to a wind turbine. The goal in buying a wind turbine is not to get the cheapest machine, but to purchase the most reliable and most durable turbine you can. Reliability and durability beat cheaper up-front cost any day.

While the governing mechanism is an important consideration, you don't have much choice. There aren't many small wind turbines on the market. And, as you may recall from previous examples, there may only be one or two—at best three—turbines that produce the amount of electricity you need. Chances are they all rely on horizontal or side furling.

Shut-down Mechanisms

Because wind turbines require routine maintenance and occasional repair, and because these operations may need to take place on slightly windy days—or winds may come up on calm days—small wind turbines should include a reliable shut-down mechanism. As the name implies, shut-down mechanisms allow a turbine to be turned off. They allow operators to maintain and repair wind turbines without fear of injury. They also provide a means of shutting a wind turbine down when extremely violent storms, especially thunderstorms, are approaching—if an operator doesn't have faith in the turbine's overspeed control.

Wind turbine designers rely on two types of shut-down mechanisms: mechanical and electrical. Let's start with mechanical systems.

Mechanical systems include disc brakes and folding tails. Disc brakes, like those found in our cars and on some more expensive bicycles, are manually activated.

> Inexpensive wind turbine designs without a reliable shut-down mechanism are a short-sighted gamble at best.
>
> Mick Sagrillo

Hurricane Protection

Even though some wind turbines have survived hurricanes, it's best to lower the turbine in a hurricane to protect it from flying debris if you have a tilt-up tower. If you can't lower a turbine to the ground, applying a brake could reduce the chances of damage.

They're attached to a cable that runs down the tower. Tightening the cable activates the brake and stops the rotor from turning. Wind generators from Wind Turbine Industries (Jacobs) incorporate cable-operated disc brakes. In these turbines, "brakes off" is the default setting. In other words, you have to tighten the cable to activate the brakes.

Although disc brakes like this may seem like a good idea, they are not fail-safe. If the cable breaks in violent storm, for example, an operator would be helpless to stop the turbine. There's no way to apply the brakes! In addition, brakes may not work if the wind speed is too high.

Some wind turbines can be shut off by manually furling the turbine. Turbines with this feature are folded with the aid of a cable attached to the base of the tower. As in the example shown in Figure 5.13, a cable winch is located at the base of the tower. It allows the operator to manually shut down a wind turbine in fierce winds or when leaving town for a while. When the tail is folded, as shown in Figures 5.14a and b, the rotor is turned sideways to the wind, greatly reducing its ability to collect energy.

Although folding the tail protects the rotor from overspeeding, it doesn't stop it from rotating. This presents a potential risk to service personnel working on the tower—unless another means of stopping the rotor, such as a disc brake or dynamic braking, is available. Furthermore, if the cable breaks in high winds, when the turbine is shut down, the tail will swing back into the wind and the wind turbine will start back up, which could damage the turbine—and injure service personnel—if the winds are strong enough.

Wind turbine designers also employ electrical brakes, known as dynamic brakes. Dynamic braking is the least expensive option and is found in a few small wind turbines, including the Xzeres Skystream 3.7 and Xzeres 442SR.

Dynamic braking is a fairly simple approach that's unique to wind turbines that use permanent

FIGURE 5.13. Cable Winch on Tower. MREA's Clay Sterling shuts down a Jacobs wind turbine by tightening the cable attached to the tail of the turbine. Credit: Dan Chiras.

magnet alternators. It consists of a switch inside the house or at the base of the tower. When the brake switch is closed, it short-circuits the wind turbine, rapidly slowing the rotor. Dynamic brakes short-circuit the three phases of the permanent magnet alternator. Electricity coming from the windings then flows back through the windings of the stator. This produces a magnetic field around the windings that opposes the magnetic field of the rotor (containing permanent magnets), slowing, even stopping, the turbine. Dynamic braking overpowers the ability of the rotor to spin the alternator.

In dynamic braking, the braking force is proportional to the rotor speed. As the rotor slows down, the braking force diminishes. As the rotor speed approaches zero, so does the braking force. In low to moderate winds, dynamic braking should either stop the rotor or slow it down considerably. Be sure to check with the manufacturer to find out if their dynamic brake completely stops the rotor in high winds. Nor do you want strong winds to be able to overpower the dynamic brake of your turbine.

Shut-down mechanisms of a wind turbine should be high on the list of considerations, right up there with swept area and top-of-tower weight. If the turbine is to be serviced *on* the tower, the shut-down mechanism should be capable of completely stopping the rotor.

Unfortunately, some small wind turbines (less than ten-foot diameter rotors) come without shut-down mechanisms to reduce manufacturing costs. While these turbines are cheaper initially, they are vulnerable to catastrophic failure. It goes without saying that replacing a wind turbine after its first thunderstorm is much more costly than purchasing and installing a turbine with a shut-down system.

Without a shut-down mechanism, homeowners are left to chance. A violent storm may destroy their turbine and they're powerless to do anything about it except maybe lower the turbine to the ground. (That's not a good idea in a strong wind!) Mick's heard many horror stories about small wind turbines that were destroyed in strong winds while the homeowners looked on helplessly. Most of them had no idea that their inexpensive wind turbine could be irreparably damaged in a storm until after the fact.

FIGURE 5.14. Homebrew Wind Turbine. Dan and his friend Pete Veronesi just completed work on this amazingly quiet, efficient, and durable wind turbine designed by the folks from Other Power in northern Colorado. However, you should be wary of most homebuilt wind turbine designs on the market. Many are not going to produce much energy or last very long. Credit: Dan Chiras.

Sound Levels

The sound a turbine produces is also a factor one should consider when buying a turbine. Sound levels are important to homeowners and their neighbors. They may also be important to zoning officials, too, as you shall see in Chapter 10.

All residential wind turbines produce sound. There's no such thing as a perfectly quiet wind turbine, although some get quite close.

Sounds emanate from five possible sources. First, when the blades spin, they produce a swishing sound. Second, when furling, the blades may produce a helicopter-like beating sound. (This occurs when a blade passes through the wake of a preceding blade.) Third, rotation of the rotor in the alternator also produces sound, depending on the controller design. Fourth, in gear-driven wind turbines, movement of the gears also produces sound. A fifth, and rather complicated source of sound, is generated in the alternator of some wind turbines. For the technically inclined readers: "It is caused by the interaction of the alternator and the inverter," says NREL's Jim Green. "The high switching frequencies in the inverter distort the current and the voltage in the alternator. This, in turn, vibrates the stator and generates sound."

Certified wind turbines list sound levels at standard conditions—5 m/s wind speed, 60 meters from the turbine.

Sound levels increase as wind speed increases. However, sound from a wind turbine is often difficult to detect and is rarely a nuisance one could label as noise. Why?

As I have pointed out previously, wind turbines are typically mounted fairly high off the ground—typically 80 to 120 feet—to reach the smoothest, most powerful winds. Because sound decreases with the square of distance, tower height significantly reduces sound levels on the ground. In addition, sound from wind turbines is often drowned out by environmental sounds. That's because, background sounds increase as wind speed increases, too. When the wind blows, leaves and branches of trees rattle and shake. Wind blowing by one's ears also makes sound. All of this background noise tends to drown out sound produced by a wind generator.

Even so, it is important to consider sound levels. One way is to listen to the turbines you are considering as they operate under a variety of wind speeds. If you can't, you may want to ask homeowners who have installed the turbines you are considering for their experiences.

Another, more scientific, method is to check out the rpm of the turbines you are considering at their rated outputs. Rated output, discussed shortly, is the output of a turbine in watts or kilowatts at a certain wind speed. Knowing how fast the blades and generator are rotating at that speed gives you an idea of the potential for sound.

Generally, the higher the rpm, the more sound the blades will make. Bear in mind, though, that it is the *tip* speed, not just the rpm, that affects the amplitude of the sound. A small turbine at 1,000 rpm could be quiet, and a much larger turbine at 50 rpm could be very loud.

Of the turbines we've been considering, the Xzeres 442 (formerly ARE442) and the Xzeres Skystream 3.7 rotate at 150 and 350 rpm, respectively, in an 11 meter-per-second wind. The Bergey Excel 10 spins at 400 rpm, the Luminous Whisper 200 rotates at 1,200 rpm, and the Luminous Whisper 500 spins at 800 rpm in wind blowing at 11 m/s. All of these turbines, except the Whisper 200 and 500, should be pretty quiet.

While the rpm of a wind turbine influence sound levels, they also give an indication of quality. Generally, less expensive and less durable turbines spin at a higher rpm. They rely on less expensive generators operating at high speeds to produce energy. In addition, higher rpm turbines are subject to more wear and tear and tend not to last as long.

Other Considerations

Although swept area, top-of-tower weight, rated output at 11 m/s, annual energy output at 5 m/s, governing mechanisms, shut-down mechanisms, and sound levels are the most important factors to consider when buying a wind turbine, there are a few others—details that manufacturers provide. Readers should be familiar with them, but not let them overly influence their judgment. They're nowhere near as important in the final analysis.

Rated Power

When we talk about wind turbines, we typically refer to them by their rated power. For example, you may be considering a 6-kW wind turbine or a 10-kW turbine. Rated power is the output in watts at rated wind speed. Rated power is determined at 11 meters per second, or 24.61 mph, as noted earlier.

Much like the output of solar-electric modules, rated power of wind turbines give customers a way to compare products. That said, rated power still isn't that useful. It is only one point on the power curve, and it's the output at a wind speed— the rated wind speed—that's much higher than those typically encountered at most sites. We're more interested in output within the typical range of wind speeds at our site, which is well below the rated speed. My advice on rated power: forget it. Select a wind turbine that has a good annual energy output at the average wind speed in your area.

Cut-in Speed

One factor that is frequently advertised is cut-in speed—the wind speed at which a wind generator starts producing electricity. It's often used as a selling point. The manufacturers argue that because the turbine starts spinning and making electricity at a lower speed, you will get more energy from your turbine. They're especially keen on this strategy in low-wind areas. They claim that low cut-in speed turbines make better use of a meager wind resource. Unfortunately, low cut-in speeds are pretty meaningless. Here's why: Although the blades of most small wind turbines may start turning in low winds, most turbines don't produce measurable amounts of electricity until wind speeds reach eight to ten miles per hour.

Consider an example: Let's compare the output of a turbine in a five mile-per-hour wind to a turbine in a ten mile-per-hour wind. Knowing that power available from the wind is a function of velocity cubed, we can compare the power in the wind by cubing five (don't worry about units) and cubing ten. When we compare $5 \times 5 \times 5 = 125$ to $10 \times 10 \times 10 = 1,000$, you can see why low wind speed cut-ins are irrelevant. Winds in the 10 to 20 mile-per-hour range are the ones we want to capture. In the course of a year, wind speeds between 10 and 20 miles per hour deliver the bulk of the annual energy output. Wind speeds below ten miles per hour contribute only marginally. Any company that uses start-up speed or very low wind speed performance as an advertising point does not understand how wind works. And, their turbines are probably pretty lightweight, and hence very likely not very durable.

Power Curves

Another intriguing but ultimately useless bit of information provided by wind turbine manufacturers is the power curve, an example of which is shown in Figure 5.11. Power curves are graphs that show the power production of a wind turbine in watts at different wind speeds.

While power curves are impressive and make for good visuals, they aren't very useful. The main reason, as shown in Figure 5.11, is that wind turbines typically don't operate at the peak of the power curve, they operate at the low end of the power curve in areas with average wind speeds of 5 to 7 meters per second—approximately 11 to 15 miles per hour. When shopping for a turbine, what's important is how much energy a wind turbine will produce at *your* site on *your* tower at *your* average wind speed—that is, once again, AEO. Your best bet is to buy a wind turbine that's been SWCC (Small Wind Certification Council) certified—one that's been field-tested.

Final Factors

When shopping for a wind turbine, be sure to check into the company's customer service record—how well it supports its dealers and customers. Some companies have notoriously poor customer service. Others, like Bergey WindPower, have stellar records.

Another factor to consider is how long the company has been in business. As a general rule, it's a good idea to do business with companies that have been around for a long while. (Bergey has been in business since the early 1980s.) It's also a good idea to purchase turbines that have been on the market for a while, so you know all the bugs have been worked out of them. It's sad but true; some manufacturers start selling wind turbines before they've been extensively field-tested. As a result, the early consumers become the beta testers.

As a rule, I recommend against buying a turbine in its first year or two on the market. Many people who made this mistake have been extremely disappointed.

When shopping, it's also important to check on the availability of parts. Parts shipped from foreign locations, such as China, Europe, or South Africa may take months to arrive, if they're available at all. Meanwhile, your $30,000 wind energy system sits idle. If you buy a turbine imported from another country, be sure the local distributors carry spare parts.

Speaking of parts, when shopping for a turbine, it's a good idea to consider the number of moving parts and wear points in a turbine. Some designs, like the Jacobs turbines manufactured by Wind Turbine Industries in Minnesota, contain a lot of moving parts resulting in numerous wear points—up to 300. These turbines may require a lot of tinkering at 100 feet. They're not for the acrophobic. Fortunately, most modern small wind turbines have many fewer moving parts and, as a result, they tend to be more reliable—provided they are built from sturdy materials.

While you are at it, be sure that the company you buy from offers technical advice should you have trouble installing the turbine or if the turbine breaks down. And do their technical support staff speak your native tongue? Although many French and Spanish turbines are tried and tested by companies that have been in the business for a long time, if you live in North America you may still have language and distance barriers to overcome. (Some foreign turbines are imported and may be reliably supported by importers.)

While you are at it, check into the warranty. Warranties typically run five years, although there are some wind turbines that come with ten-year and many that only come with two-year warranties. I'd be wary of any turbine with a warranty shorter than five years. Warranties often reflect product longevity. If a manufacturer won't guarantee a turbine for more than two years, chances are you will start having trouble with it once the warranty expires. With wind turbines, the rule is, the longer the warranty, the better. Also be sure to determine what the company covers. Is it materials and workmanship or parts only? Does it include shipping? Unfortunately, most warranties do not cover the cost of labor.

When buying a wind turbine for a grid-connected system, be sure that the inverter is compliant with Underwriter's Laboratory requirements (UL 1741) or its equivalent. This ensures safe interconnection with the grid. More specifically, it ensures that the inverter will shut off and not backfeed onto the grid should there be a power outage. While lack of UL 1741 certification is not an indication of a lack of safety, it will cause you problems when you try to get approval to connect your system to the utility.

If you are not up to installing the system yourself, be sure to find a local installer who will do it. That way, you'll have local support. Be sure to ask for references and interview former customers.

If you buy from an online wholesaler and install the system yourself, be aware that most wholesalers don't offer technical support, installation advice, or assis-

What Voltage?

As you shop for wind turbines for wind systems with battery backup, you'll find that they come in various voltages—typically, 12, 24, and 48 volts. Which voltage is best for your application?

As a rule, we recommend higher-voltage wind systems for modern homes. (Lower voltages may be appropriate for small cabins and boats.) Higher-voltage systems are more efficient, especially if long wire runs are required. They are more efficient because higher voltage results in lower current, and lower current results in lower losses in the wire. (Wire losses are a function of the amount of current in a wire. The higher the current, the higher the losses. The lower the current, the lower the losses.) Current can be decreased by increasing the voltage.

Why does an increase in voltage decrease the current through a wire?

This can be explained by referring to the equation: watts = amps x volts (sometimes called the *power formula*). If 12 amps of electricity are flowing at 12 volts, the wattage is 144. If the voltage is increased to 24, and the wattage remains the same, the amperage drops to 6. If the voltage is increased to 48, and the wattage remains the same, the amperage drops to 3. As a result, a 48-volt system can use much smaller wire than a 24-volt system. A 24-volt system can use much smaller wire than a 12-volt system. For a mathematical explanation of this phenomenon, see the sidebar "Current and Wire Losses."

Current and Wire Losses

Voltage drops over long wire runs. This phenomenon is due to resistance created by the conductor (usually copper) in the wires running from the turbine to the controller. Wire losses are a function of the resistance a wire poses to electrical current. Resistance is a function of run length and the size of the wire. It is also affected by the type of conductor. Copper, for instance, is a better conductor than aluminum—it poses less resistance. As a rule, it is best to keep wire runs as short as possible and to use larger gauge wire to lower resistance and improve the efficiency of your wind system.

For the mathematically inclined, wire losses are proportional to the resistance of the wire (just described) and the square of the current. Here's what's interesting about current: If you double the voltage, the current is cut in half and wire losses are reduced by a factor of four. For long wire runs, higher voltage also means that you can avoid using expensive heavy-gauge wire.

tance. They may also not offer replacement parts or repair services. If you buy cheap, you are on your own when you need help.

All in all, I believe that it is a good idea to stick with name brand wind turbines and avoid newcomers and foreign imports, especially Chinese- or Mexican-made wind turbines—at least at this time.

Buying a used or reconditioned wind turbine may be an economical option, and will typically cost about 50 percent to 60 percent less than buying a new turbine—potentially saving you a huge amount of money. The problem with used wind turbines is that you don't always know what you're getting and, if you did, you wouldn't be happy. "Used machines from early California wind farms," notes small wind energy expert Paul Gipe, "have been beat to death." "eBay is not a good place to buy a used turbine that works," notes Aaron Godwin. If the turbine is not reconditioned, it may cost a fortune in time and parts to rebuild it. Only the foolish install a non-rebuilt wind turbine, and they usually only make that mistake once.

If you buy a remanufactured wind turbine, like a Jacobs 31-20 or the larger Vestas, from companies that refurbish these workhorses, "be sure that the machine has been fully reconditioned, not just painted and had new blades installed," advises Godwin. Also, when buying one of the larger reconditioned wind turbines (25-foot and longer blades), be sure that the controller or inverter that comes with the unit meets current standards for utility connection. You may also want to obtain an extended warranty. Also be sure you can find a tower for the turbine *before* you purchase it. Suitable tall towers for the larger small-scale wind turbines are hard to find.

Although a few good deals on used turbines are available from time to time, I recommend extreme caution when considering this route. Shop very carefully. When it comes to issues of maintenance, performance, and the reliability of a used turbine, you have only the word of the seller to go on.

Building Your Own Wind Turbine

If you would like to tap the power of the wind, but don't have the cash to buy a system, or you love to tinker and want the challenge of making your own turbine, you may want to consider building your own wind turbine (Figure 5.15). Building your own wind generator provides invaluable experience—and helps you appreciate all of the hard work that has gone into residential wind turbine design over the years.

As readers likely suspect, the internet is full of YouTube videos, schematic diagrams, and written directions on building a wind turbine. You will find many sets of instructions on ways to build small horizontal and vertical-axis wind turbines. Simply search for "build your own wind turbine."

As you search the internet, you will find wind turbines made from the DC motors from treadmills with blades from commercially available microturbines. You will find homemade wind turbines made from car alternators with blades carved from 2 × 4s. You will even find turbines made from PVC pipe. Be careful. Use common sense and your knowledge of wind turbines to select a design. PVC wind turbines may work for a while, but they are likely to break and fall apart after a while or in strong winds.

If you are serious about wind energy, and want to build a turbine that produces a significant amount of electricity, think big. Many of the designs are just too small to produce much energy.

In your search, you will very likely encounter an article by Steve Hicks, entitled "Building Your Own Wind Generator." It's available through the American Wind Energy Association's website and is well worth reading. There are even a couple of books on the subject: *A Wind Turbine Recipe Book* and *How to Build a Wind Turbine* by wind expert and inventor Hugh Piggott and *Homebrew Wind Power: A Hands-On Guide to Harnessing the Wind* by wind experts Dan Bartmann and Dan Fink. The latter is an extremely well-written and thorough treatise on

FIGURE 5.15. This is a picture of my friend Pete Veronesi and me standing beside a homemade wind turbine that we made at a workshop in Colorado.

the subject. The turbines described in these books are built to last and to produce a lot more energy than the garage-tinkerer designs that flood the internet.

You can also find information on these authors' websites: Hugh Piggott's scoraigwind.com and otherpower.com. These sites offer a wealth of valuable information, including links to other websites. They also offer plans on building axial-flux alternator turbines.[3] Their turbines were inspired by Hugh Piggott.

Rather than provide a full description of this subject, which would take a chapter or two, maybe even an entire book, I recommend you look at these and other resources—always exercising your critical thinking skills. I'll make a few salient points and let you decide if you want to build your own turbine.

Building a wind generator is not that difficult, if you are mechanically inclined. You will need a shop and some common hand and power tools. You may need a welder and a cutting torch, as well as a grinder. As a rule, wind turbines with 12-foot or larger blades are beyond the abilities of most home builders, as they require very strong construction skills. Smaller turbines, with 8–11-foot blades, are easier to build.

"The hardest part of designing a small windmill for electricity production is to find a suitable generator," writes Piggott in his book *Windpower Workshop*. "For best performance you'll need a reliable, low-speed generator that's pretty efficient in light winds," he adds. Piggott strongly recommends permanent magnet alternators. "They are," he says, "by far the most popular choice on successful small windmill designs." Their alternators do not require brushes or slip rings. The only parts that can wear out are the bearings.

Unfortunately, permanent magnetic alternators are difficult to find. Piggott lists some sources, among them motorcycle alternators and welders. Unfortunately, each of those has a significant downside. Motorcycle alternators, for instance, are designed for high-rpm performance and are not easily adaptable to direct drive wind turbines.

Another option, says Piggott, are brushless DC motors. Brushless DC motors are like permanent magnet alternators in many ways and are an almost perfect choice. They're currently used in some new washing machines, power tools, and medical equipment.

Car alternators are popular among do-it-yourselfers. They are widely available and are fairly inexpensive, even if purchased new. They are also designed to charge batteries. Unfortunately, car alternators require high rpms and are not always very efficient. They require a lot of electricity to power the electromagnets that create the magnetic field. Moreover, if you go this route, you'll need to rewind the stator coils (these are the coils that produce electricity) to at least double the number of windings. This enables the alternator to produce electricity at the lower speeds common in wind turbines. Even then, the power output curve of a car alternator is poorly matched to the power curve of a wind turbine rotor, resulting in poor performance in low winds and out-of-control rpm at higher wind speeds. "Most people who have tried adapting car alternators to wind turbines have given up in frustration. Their underperforming wind turbines are strewn across the countryside after the first good storm," notes Mick Sagrillo.

Building a generator is challenging and vital to the success of a homemade wind turbine. But it's not all you'll have to do. You'll have to design and build blades and install controls.

Blades and rotors are particularly tricky. Because they are subject to considerable stress, they need to be well constructed. "For minimum stress, the blade should be light at the tip and strong near the root," recommends Piggott.

Most homemade wind turbines use wooden blades. Wood is light, strong, and readily available. It is also easy to fashion into a workable blade and resistant to stress. Metal, on the other hand, is prone to stress, and steel is heavy. Fiberglass is great, but most homeowners don't have the equipment or expertise needed to mold fiberglass blades.

When designing blades, you'll need to select the rotor diameter very carefully. The diameter should match the operating speed of the alternator. Piggott's book and website provide excellent advice on the subject, along with advice on carving, painting, and balancing blades.

Another option is to purchase a set of blades made for a commercially produced small wind turbine. This could save a lot of work, but blades are not generally cheap. And, they may not be suited to the generator you are using.

A rotor and a generator do not a wind energy system make. You'll also need to devise a frame to mount the generator and tail vane and install necessary controls. Overspeed controls and shut-down controls, discussed in this chapter, should be incorporated, if at all possible. Otherwise your wind turbine could be destroyed by the first storm that passes through your area. A charge controller (discussed in Chapter 3) should also be installed in battery-based systems to prevent batteries from being overcharged. You'll also have to install a dump load, a resistor to dump excess electricity when the batteries are full.

As with any wind turbine, you'll also need to install the turbine on a suitable tower. Tower designs can also be found online. Be careful, however; homemade towers often leave much to be desired. Towers are far more complicated to design than wind turbines. Your best and safest bet is to buy a used tower, if you want to save money. And of course, you need to install it properly so it is safe.

Building your own turbine is fun and challenging, and can save a lot of money. Moreover, there's lots of good advice out there on the subject. And, although I'd never discourage anyone from giving it a try, I'd be remiss if I did not point out that homemade wind turbines may not produce as well as wind turbines sold by any of the world's best manufacturers, such as Bergey, Ventera, and Xzeres.

The best manufacturers design, engineer, and build their small wind turbines with performance and reliability in mind. Companies like Bergey WindPower have been relentless in their pursuit of quality and continue to improve their products, making the most reliable and efficient wind generators humanly possible. Homemade wind turbines can't match these highly refined, finely tuned turbines, but they're a lot of fun to build and may even provide some of your electrical needs! The only exceptions that I have seen are the designs available from Hugh Piggott and Other Power, mentioned earlier. Their turbines run extremely quietly and produce energy on par with manufactured small wind turbines. Even though they may not be as technologically sophisticated as modern wind turbines, well-made home-built wind turbines like these can last for many years if properly maintained, and sometimes survive and perform better than light or medium-duty manufactured turbines. And if you build your own, you'll really understand your wind generator!

Conclusion

By now, you realize that the simple wind turbine you've been eyeing on a neighbor's property is a rather complicated and fine-tuned turbine. Moreover, you may have discovered that buying a wind turbine won't be as easy as you may have thought. There are a lot of factors to take into account. Price is only one of them. Be sure to choose a turbine that meets your needs. If you're installing a grid-connected system, choose a turbine that is designed for utility interconnection. If you're installing a battery-based system, you'll need a battery-charging turbine.

Also, take into consideration key factors like the swept area of the turbine and the annual energy output at the wind speed on the top of the tower at your site. If you want a good turbine, one that will last for many years, you'll need to look for one of the heavier turbines. They will cost more initially, but will outlast their cheaper cousins, costing much less in the long run.

CHAPTER 6

Towers and Tower Installation

Many people considering wind energy focus their attention on the turbine and, therefore, pay little attention to another essential component of a successful wind system, the tower. Focusing attention on the wind turbine is quite natural. After all, it's the collector—that part of the system that captures and converts the wind's energy into electricity. Sitting high atop their towers, spinning in brisk energy-rich winds, turbines are the superstars of wind systems.

As any owner of a successful wind system will tell you, however, it is not just what's on top of the tower that counts. The tower also plays a huge part in creating a successful wind energy system. The tower not only raises a wind turbine to a height at which it can harvest lots of energy from the wind, it also is built to withstand Nature's fury. In addition, the tower can be quite costly. In large systems, the cost of a tower may account for nearly a quarter of the system cost. For smaller wind

Is Wind Right for You?

Periodic inspection and maintenance and occasional repair of wind turbines are essential to the long-term success of a wind energy system. The towers on which they stand present a formidable barrier to these activities. Many wind system owners fail to perform these tasks because they don't want to lower or climb their towers once a year. So, if you are a "put it up and forget about it" kind of person and can't afford to hire someone to perform an annual inspection and maintenance,

you may want to consider installing a solar-electric system instead. Batteryless solar-electric systems are as close to maintenance-free technology as you can get. If you install a wind system, you will either need to climb the tower or lower it to the ground at least once a year (some experts recommend twice a year) to inspect the turbine, wires, connections, and perform maintenance, as required. (I'll describe this process in more detail in Chapter 9.)

turbines, an appropriately sized tower plus installation may cost two to three, sometimes five times, more than the turbine.

Tower installation involves a significant investment in time, money, and energy. Most of the installation time for a small wind system is spent building the foundation and anchors for the tower, assembling, and raising the tower.

During a wind energy system installation, it can take one to four days to excavate and build a foundation for a tower and another two to five days to assemble it. Assembling the turbine, attaching the turbine to the tower, and making the electrical connections may only require two to four hours. System wiring, especially

A Primer on Wind Generator Tower Physics

Suppose you are getting ready to paint the kitchen in your home. You're set to go, but discover that you have forgotten the screwdriver needed to pry open the paint can lid. It's in your toolbox in the garage. It's ten degrees below zero outside and you're not feeling like braving the cold. What to do?

You could bundle up and trudge out to the garage to retrieve it, or you could fish around in your pocket for a quarter to open the paint can. Since the quarter in your pocket involves a lot less work than retrieving a screwdriver from your workbench, and you are raring to go, you opt for the quarter. What you find, however, is that you expend at least as much effort prying open the lid of the paint can with a quarter as it would have taken to retrieve a screwdriver from the garage. Why?

To open the paint can, you need a device that physicists call a lever. Both a quarter and a screwdriver can serve as levers to pry open the lid. Although both "tools" will work, much more force is required to open the paint can with a quarter than with a screwdriver. Why?

In a lever, a force is applied to one end and the lever transfers that force to some object at its other end. How the force applied to the lever compares to the force transmitted to the object depends on the position of the pivot point. If the pivot is in the center of the lever, the forces will be equal. But if the pivot is closer to the object (load end), the force at that end will be greater than the applied force.

The equation that describes this is: $f_1 \times d_1 = f_2 \times d_2$, where f_1 is the applied force and d_1 is the distance from the applied force to the pivot, f_2 is the force at the load end and d_2 is the distance from the pivot to the load end, shown in Figure 6.1. Rearranging the equation to solve for the force at the load end gives: $f_2 = f_1 \times d_1/d_2$. Since the distance from the edge of the paint can (the pivot point) to the lip of the lid (d_2) is the same for both the quarter and the screwdriver, but the screwdriver has a much longer lever arm (d_1) than the quarter, the screwdriver will deliver much more force to the lid of the can.

So, what does all of this have to do with a tower for a wind turbine? It's quite simple. It turns out for battery-based systems, accounts for another significant portion of the installation time.

Given the importance of the tower, its cost, and other considerations, it is vital that you understand a bit about towers *before* you shop for a wind turbine. In the pages that follow, I'll explore three tower options. I'll briefly describe tower assembly and installation. I'll discuss proper siting to help you understand where a tower and wind turbine should go on your property. I'll discuss the economic benefits of installing a tall tower vs. a short tower. I'll end with a description of safety concerns and give you some advice on buying a tower.

that the wind views the tower as a lever, and the wind applies a force to one end of the lever, the top of the tower. The downwind pier of the foundation, shown in Figure 6.1, is the pivot point, and the upwind pier of the foundation is the other end of the lever. In a tower where the height is ten times the distance between the downwind and upwind piers, a wind force applied to the top of the tower will be multiplied by ten. This force acts to lift the upwind foundation out of the ground.

The amount of force the wind can apply is directly related to two factors. The first is the wind speed. The greater the wind speed, the more force it applies. The second factor is the swept area of the rotor and the frontal area of the tower. The greater the area of the rotor and tower, the more force the wind applies to lifting the foundation.

"As swept area, frontal area, and wind speed (force) increase and tower height (distance) increases, so does the overturning force exerted by the wind on the tower and its concrete footings," notes Mick Sagrillo.

For a successful wind system, then, the tower needs to be strong enough and anchored to the ground well enough to thwart the wind's potentially destructive force. Never underestimate the power of the wind.

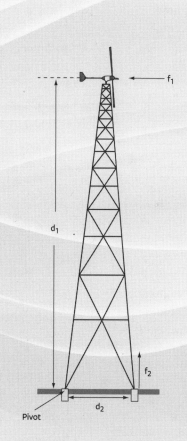

FIGURE 6.1. Forces Acting on a Tower. Credit: Bob Aram and Anil Rao.

Before we begin our exploration of the three types of towers, it is useful to discuss the forces that act on them. This will help you understand why it is important to buy—or, if you're a do-it-yourselfer, to build—a high-quality tower. It will also help you understand the importance of properly designed foundations and anchors. In order to understand this, we present an example Mick Sagrillo uses in his wind energy workshops in the accompanying textbox. It explains the forces that could pry a wind tower from its foundations.

Tower Options

Towers for small wind machines come in three basic varieties: (1) freestanding; (2) fixed, guyed; and (3) tilt-up towers. Each type has some variations, listed in sidebar. Figure 6.2 shows what each looks like.

Freestanding Towers

As their name implies, freestanding wind generator towers are self-supporting structures. They stand on their own, like flag poles or street lights, or the Eiffel Tower in Paris. Freestanding towers for small wind turbines are typically made of

Tower Types

1. Freestanding:
   a) Lattice
   b) Monopole
2. Fixed, Guyed:
   a) Lattice
   b) Tubular (typically homebuilt)
3. Tilt-up:
   a) Guyed tubular
   b) Guyed lattice

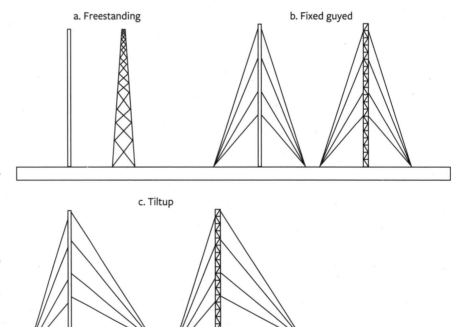

FIGURE 6.2. Wind Tower Options. (a) Freestanding towers, (b) fixed, guyed towers, and (c) tilt-up towers. Freestanding towers can be both lattice and monopoles (shown here). Fixed, guyed towers are typically lattice towers, although there are some fixed, guyed tubular towers. Tilt-up towers can be either lattice or tubular. Credit: Forrest Chiras.

steel and are firmly anchored to the ground by well-reinforced concrete foundations, discussed shortly. They do not require wires or steel cables (known as *guy cables* or *guy wires*, not *guide wires*) to hold them up. The combination of heavy-duty steel tower construction and a secure anchorage ensures that the tower can withstand powerful winds that, if they had their way, would pry the foundation loose and topple the tower. They also ensure that the tower can support the turbine.

The most common type of freestanding tower is a lattice or truss tower, like the one shown in Figure 6.3. The Eiffel Tower in Paris is a good example of a lattice structure.

Lattice towers are made from tubular steel or angle iron with horizontal and diagonal cross-bracing. The cross-bracing is typically bolted to the vertical steel legs, although it may also be welded in place. Ladders are often incorporated into the design so the towers can be climbed for inspection, maintenance, and repair. In some lattice towers, step-bolts are installed on one vertical leg for climbing.

Larger freestanding lattice towers are sometimes fitted with a small platform near the top. The platform provides a secure place to work. Nonetheless, workers should wear safety harnesses and should be secured to the tower via lanyards when working on a platform for obvious reasons—it's a long way down! (More on this safety measure shortly).

Another, albeit considerably more expensive, option for freestanding towers is the monopole: a single, sturdy pole (Figure 6.4). These sleek towers are made from round tubular steel. Rungs or foot pegs are included on the tower for climbing.

Lattice and monopole towers are secured to massive foundations made of concrete and reinforced with steel (rebar). As shown in Figure 6.5, each leg is anchored to its own foundation pier. In some designs, the piers that support the legs are connected to a huge block of concrete at their base. In a monopole tower, there's one deep, steel-reinforced concrete central foundation for support. The taller the tower, the deeper and heftier the foundation.

FIGURE 6.3. Lattice Tower. Freestanding lattice towers are made of (a) heavy-duty angle iron, as in this tower erected at the Midwest Renewable Energy Association's headquarters, or (b) tubular steel. Horizontal and diagonal steel braces made of angle iron run between the steel legs of both types of towers giving it greater rigidity and strength. Credit: Dan Chiras.

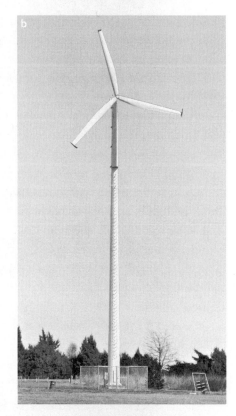

FIGURE 6.4. Monopole Tower. Monopole towers are sturdy, well-anchored by a solid foundation, but extremely costly. (*a*) This monopole tower is equipped with rungs for climbing. (*b*) Fencing is often installed to discourage unauthorized personnel from climbing the tower. Credit: Dan Chiras.

FIGURE 6.5. Concrete Piers and Base of Tower. This massive, deep foundation was built to support a 120-foot freestanding lattice tower. Each leg of the tower attaches to a steel leg embedded in each of the piers. The piers and base of the foundation are made of concrete reinforced with rebar. Credit: Mick Sagrillo.

Assembling and Installing Freestanding Towers

Freestanding lattice towers are typically assembled on the ground—one section at a time (Figure 6.6a). The Jacobs 31-20 tower the Midwest Renewable Energy Association installed in the fall of 2007, for instance, was assembled on the ground in 20-foot sections. The legs and cross braces were bolted to the legs using stainless steel nuts, bolts, and washers. Each nut was tightened using a torque wrench.

After a lattice tower is assembled on the ground, the wind turbine is often attached. The tower and turbine are lifted with a crane. The tower is then bolted to steel anchors embedded in the concrete foundation. In some instances, the tower is erected without the turbine. The turbine is hoisted onto the top of the tower with a crane once the tower is secured.

To facilitate tower construction, some lattice towers are hinged at the base (Figures 6.6b,c). This allows workers to assemble the tower on the ground, then tilt it up into position using a crane.

Cranes must have a lifting capacity sufficient to handle the tower (or the combined weight of the turbine and tower—if both are lifted at the same time). They must also have sufficient height and reach (Figure 6.7). We used an 80-ton crane to lift the 120-foot tower and Jacobs 31-20 turbine, weighing in excess of 5 tons, at the MREA headquarters, but we still needed a heavy-duty industrial forklift to assist (Figure 6.8).

A reasonably level area is also needed to assemble a freestanding tower and to lift it with the crane. Care must be taken when initially lifting the tower to prevent the blades of the turbine from breaking. This is accomplished by rotating the turbine so the hub faces the ground. (That way, it won't swing around and snap the blades as the tower is lifted.)

FIGURE 6.6. Hinged Lattice Tower. (*a*) This lattice tower is assembled on the ground but (*b*) hinged at the base. Once the tower is assembled and the turbine is attached, the tower is tilted into position using a crane, and (*c*) the hinge is bolted shut. Credit: Dan Chiras.

FIGURE 6.7. Crane Lifting Tower and Turbine. This 80-ton crane lifts a massive turbine and tower into place. The hinges at the base of the tower allow the crane to tilt the tower into position. Credit: Dan Chiras.

Two workers hold the blades in place as the turbine is initially raised.

Although freestanding lattice towers are typically assembled on the ground and lifted with a crane, it is possible on inaccessible sites to construct towers vertically one section at a time using a vertical gin pole. This technique is time-consuming and requires extreme caution.

Like lattice towers, monopole towers come in sections. They are fitted together on the ground. When completed, the tower is hoisted into place with a crane. The tower is anchored to the ground by a deep steel-reinforced concrete foundation.

Pros and Cons of Freestanding Towers

Freestanding towers offer advantages over towers supported by guy cables. One of the most important is that they require much less space (Figure 6.9). Their smaller footprint makes freestanding wind generator towers ideal for locations with extensive tree cover—although you will still need to be able to get a crane on site. If space is at a premium, give the freestanding tower serious consideration.

Freestanding towers are more aesthetically appealing to many people than guyed towers. A freestanding tower is also one of the safest towers to install, if you

FIGURE 6.8. Heavy-duty Industrial Forklift. To assist the crane when lifting a massive turbine and tower into place, you may need a little help from one of these guys. It was also used to hold the top leg in place when assembling the tower on the ground so bracing could be installed. Credit: Dan Chiras.

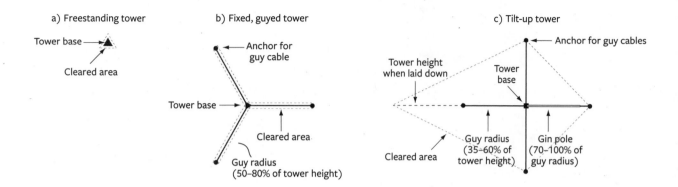

a) Freestanding tower

Tower base

Cleared area

b) Fixed, guyed tower

Anchor for
guy cable

Tower base

Cleared area

Guy radius
(50–80% of tower height)

c) Tilt-up tower

Anchor for guy cables

Tower height
when laid down

Tower
base

Cleared area

Guy radius
(35–60% of
tower height)

Gin pole
(70–100% of
guy radius)

FIGURE 6.9. Footprint of Wind Generator Towers. (*a*) Freestanding, (*b*) Fixed, guyed, and (*c*) Tilt-up. Credit: Ian Woofenden and Anil Rao.

know what you are doing, of course. Almost all the work can be done on the ground, and a single crane lift can erect the tower, turbine, wiring, etc. However, after the tower is bolted to the foundation, one worker needs to climb the tower to release the crane attachment.

Because of the large amount of steel and the need for solid (deep) concrete foundations, and the labor to install them, freestanding monopole towers are pretty expensive—too pricey for most customers. A 120-foot freestanding monopole, could cost $50,000 to $60,000 to install. Freestanding lattice towers cost less, about $17,000 to $20,000 plus $10,000 to $12,000 to install—so, somewhere between $27,000 and $32,000, depending on a number of factors.

Freestanding towers are so much more expensive than guyed towers because they require a considerable amount of steel and concrete to withstand the wind's overturning forces. In contrast, all guyed towers use trigonometry to their advantage. In a guyed tower, the guy cables shift the loads exerted on the tower base away from the tower to the much less expensive—but very strong—cables. As a result, the amount of steel required in guyed towers is much less than in freestanding towers.

Because they require so much concrete and steel and because these materials require huge amounts of energy to produce, freestanding towers also have a much higher embodied energy than other options. Embodied energy is the energy that it takes to make a product—from the extraction of the raw materials to the completion of the finished product, including shipping to retail outlets where it is sold. If your primary motivation is to decrease your environmental footprint by using renewable energy, it's instructive to think twice about how much energy it takes to manufacture and install various types of towers.

Another potential downside of freestanding towers is that they require periodic ascent to perform routine inspection, maintenance, and repair. To prevent catastrophic falls, a safety harness or safety work belt must be worn (Figure 6.10). Towers should be equipped with a safety cable that runs the length of the tower along the climbing rungs or ladder (Figure 6.11). Workers attach their safety harness to the cable when climbing with a device known as an *anti-fall mechanism*, such as a Lad-Saf, one commercially available product. It allows the worker to ascend and descend the tower, but it locks if the worker suddenly loses his or her footing and falls.

As shown in Figure 6.10, safety harnesses comes equipped with several D-rings. They're used to secure a worker to the tower via anti-fall mechanisms and lanyards to prevent falls when working on a tower. A "positioning" or "restraint" lanyard holds a worker in place to allow him or her to work hands-free. A "shock absorbing" lanyard arrests a fall, slowing the worker gradually to prevent a harmful jerk.

Even though good safety equipment prevents injury, many people don't like the idea of climbing a tower. If you are not willing or able to climb a tower, you must be willing to hire someone to do it. If not, consider installing a tilt-up tower or a solar-electric system.

Fixed, Guyed Towers

The second major type of tower is the fixed, guyed tower (Figures 6.12 a and b). They are often used to mount radio antennas for ham radio operators. Most are lattice towers, and most lattice towers consist of triangular lattice sections that come in 10- or 20-foot sections. The legs of fixed, guyed towers are made of steel tube or pipe, or sometimes solid steel rods. The three legs of the lattice tower are usually 18 inches apart and are secured by horizontal and diagonal steel cross braces (Figure 6.12b). Unlike freestanding lattice towers,

FIGURE 6.10. Safety Harness. Mick Sagrillo demonstrates proper use of a safety harness in one of his workshops. Credit: Dan Chiras.

FIGURE 6.11. Safety Cable and Lad-Saf. (*a*) Worker prepares to climb tower. Note safety harness and Lad-Saf attached to safety cable. This prevents the worker from falling. (*b*) Close-up of Lad-Saf and safety cable. Credit: Dan Chiras.

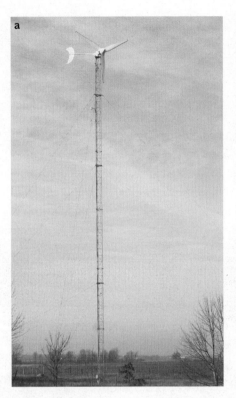

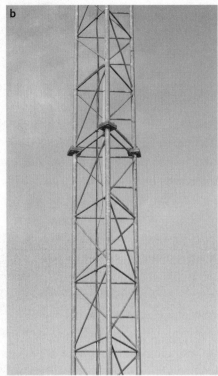

FIGURE 6.12. Fixed, Guyed Lattice Tower. (*a*) This lattice tower is anchored by guy cables (barely visible in photo) and is one of the most popular and least expensive tower options. (*b*) Close-up showing details. Credit: Dan Chiras.

the legs of a guyed lattice tower do not taper from base to top. To inspect and maintain the tower and turbine, you'll need to climb the tower.

Guyed towers are bolted onto a concrete foundation (base) and are supported by guy cables (Figure 6.12a). Guy cables are made of high-strength stranded-steel cable. They extend from attachments on the tower to anchors embedded in the ground. Guy cables are strung out in three directions and are anchored 120 degrees apart. The guy radius, that is, the distance from the base of the tower to the anchors, is usually about 75 percent of the tower height, but that distance can range from 50 to 80 percent, depending on the construction of the tower. For a 120-foot tower, then, the anchors would be 120 degrees apart and 60 to 100 feet from the base of the tower. (Be sure to check with manufacturer specifications or consult with a qualified structural engineer.)

Fixed, guyed towers are also made from pipe or tubular steel which comes in 20-foot sections. These are either slip-fitted together with steel couplings or screwed into steel couplings. Like guyed lattice towers, tubular towers are supported by guy cables.

Assembling and Installing Fixed, Guyed Towers

Guyed towers are usually assembled on the ground. Lattice towers are bolted to-gether. For tubular towers, one section of pipe slips into or screws into a coupling attached to the next section.

After the tower is assembled, the turbine and guy cables are attached and the tower and turbine are erected by a crane. In some cases it may be necessary to lift a lattice tower or tower made from steel tubing in smaller sections, depending on its height and strength. That's because some towers are rather flexible and may not be strong enough to permit the tower to be lifted all at once. This is typically encountered when the tower is 80-feet or longer. In such cases, the turbine is lifted onto the tower after the last section is in place.

Fixed, guyed towers can be assembled vertically, one section at a time, using a device known as a *vertical gin pole*. This type of gin pole is an inexpensive and temporary vertical crane. It is bolted onto the tower and is used to raise sections of the tower one at a time. After a section is in place, the gin pole is moved up so the next section can be then installed, and so on. Vertical gin pole assembly is time-consuming and tedious and can be a bit dangerous. Those who've tried it, often recommend against it. If no crane is available or the crane cannot access the site, however, a vertical gin pole may be your only option.

Fixed, guyed towers are supported by concrete foundations, though they are generally not bolted to them. Guy cables are attached to the tower prior to erec-tion and attached to anchors, usually steel-reinforced concrete. Once the tower is upright and plumbed, workers tension the cables. If the turbine was not previously attached, it is then lifted by the crane and fastened to the top of the tower.

Pros and Cons of Fixed, Guyed Towers

Fixed, guyed towers cost much less than freestanding and tilt-up towers. They're generally the least expensive tower option, which is why they are so common. Lower costs are due to the fact that guyed lattice towers require much less steel and their foundations require a lot less concrete than those of freestanding towers. Lattice towers used by installers are also mass produced for the telecommunications in-dustry, making them less expensive and widely available. (If you shop around on Craigslist you may be able to find used tower sections that are priced much cheaper than new ones.) Another advantage of fixed, guyed towers is that they require less space than tilt-up towers, discussed next.

On the downside, fixed, guyed towers must be climbed to perform routine inspection, maintenance, and repair on the turbine and tower, like freestanding towers. Moreover, fixed, guyed towers require more space than freestanding towers, as you saw in Figure 6.9. Another downside is aesthetics. Some people view the guy cables as an eyesore. Up-close, the numerous guy cables are pretty ugly, however, I should point out that the cables "disappear" into the background with distance, as you can see in Figure 6.12. The guy wires on my 126-foot tower are visible one-tenth of a mile away, but are not very obtrusive. Two tenths of a mile away, you can still see them, but barely. Three tenths of a mile away, and I can't see them at all.

Guy cables may also present a hazard to birds. "Birds can fly into the wires by accident because they are small diameter and difficult to see," NREL's Jim Green points out, although such occurrences are rare.

Tilt-up Towers

The third type of tower option is a guyed tilt-up tower. As their name implies, guyed tilt-up towers are supported by guy cables. In addition, these towers can be raised and lowered (tilted up and down) to allow you to perform inspection, maintenance, and repair.

Guyed tilt-up towers come in many heights— up to around 130 feet (40 meters). Historically, most tilt-up towers (like mine) have been made from steel pipe, although guyed tilt-up lattice towers are also available through Bergey WindPower.

Guyed tilt-up towers require four sets of guy cables at each level. Cables are located 90 degrees apart. As in the fixed, guyed tower, guy cables of guyed tilt-up towers provide support. The fourth set of cables, however, enables workers to safely raise and lower the tower. Without the additional set of cables, the tower would topple whenever it is raised or lowered.

As illustrated in Figure 6.13, a tilt-up tower is raised and lowered with the aid of a gin pole. Unlike the vertical gin pole used to erect a tower, the gin pole in guyed tilt-up towers is permanently attached to the mast (the pole or lattice tower). This steel pipe (or a section of lattice tower) is permanently attached to the base of the tower at a right angle (a 90° angle) to the mast. It serves

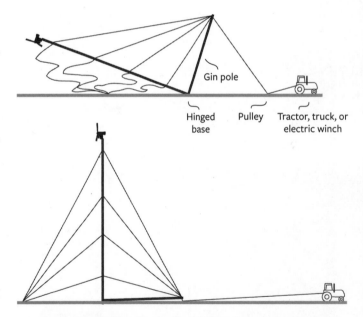

FIGURE 6.13. Guyed Tilt-up Tower. Guyed tilt-up towers are raised and lowered using a truck, tractor, electric winch, or grip hoist. By far the safest is an electric winch. You typically only need a ½- to 1-horsepower electric motor. I'd avoid trucks or tractors, as raising and lowering a tower should be slow and steady. Credit: Anil Rao.

FIGURE 6.14. Hinged Base Guyed Tilt-up Tower. The hinge at the base of this tilt-up tower allows it to be tilted up and down to maintain and service the wind turbine. The hinge is anchored to a concrete foundation. Credit: Dan Chiras.

as a lever arm that allows the tower to be tilted up and down for inspection, maintenance, and repair.

Tilting a tower up and down is also made possible by a hinge anchored to the concrete base (Figure 6.14). This hinge forms a joint between the foundation and the base of the tower that operates much like an elbow joint. When the tower is down—that is, lying on the ground and ready to be raised—the gin pole sticks straight up. When the tower is vertical, the gin pole lies near and parallel to the ground. As illustrated in Figure 6.13, a steel cable connects the free end of the gin pole to a lifting device, preferably an electric winch, which is used to raise and lower the tower, as discussed shortly.

As in fixed, guyed towers, guy cables hold a tilt-up tower upright and resist the force of the wind. Guy cables typically connect to each section of tower. The taller the tower, the more sets you'll need.

As with fixed, guyed towers, guy cables attach to anchors. The guy radius is 35 to 80 percent of tower height. For a 120-foot tilt-up tower, then, the anchors should be located 42 to 96 feet from the base of the tower, depending on manufacturer's recommendations. All but one of the fourth set of guy cables is permanently attached to the end of the gin pole, as shown in Figure 6.15. The free end of the gin pole is connected to the fourth anchor. As shown in Figure 6.15, the uppermost guy cable is often connected to its own anchor as an added measure of safety.

FIGURE 6.15. Gin Pole. The gin pole on The Evergreen Institute's tilt-up tower lies parallel to the ground when the tower is erect. It is attached to an anchor located at the end of the pole. Notice that the guy cables for all but the top section are connected to the end of the gin pole. The uppermost guy cable is attached to its own anchor as an added measure of safety. Credit: Dan Chiras.

Assembling and Raising a Tilt-up Guyed Tower

The steel pipe or tubing typically used for guyed tilt-up towers comes in 21-foot lengths. The individual lengths are secured by bolts or joined by slip-fit couplings on the ground. While the tower is on the ground, guy cables are attached to the tower and the concrete anchors.

Once assembled, the tower is tilted into position. This is accomplished with the assistance of a tractor, a pickup truck, a heavy-duty electric winch, or a manually operated grip hoist. The lifting device is attached by a cable to the end of the gin pole. Although they're slow compared to other options, grip hoists are safe and affordable (Figure 6.16). "They provide a level of control that no other lifting method offers," says Ian. "Faster methods can lead to disaster before workers notice a problem. Slow and steady sometimes is best."

When installing a tall tower for the first time, some installers raise one or two sections of the tower at a time (Figure 6.17). After one section is raised, the tower is plumbed and the guy cables are properly tensioned. This section is then lowered so that an additional segment can be added. It is then raised, plumbed, and cables tensioned. This process continues until the entire tower is assembled, plumbed, and properly tensioned.

Experienced installers also recommend lifting (and plumbing) the entire tower *before* attaching the wind turbine just to be sure that everything is working. As Ian notes, "Tilt-up towers *will* be 'noodley' and out of plumb when first raised. You don't want the added weight of a wind turbine to deal with. And you don't want to risk damaging a wind turbine should the tower collapse because it was not plumbed or the cables were not tensioned properly, or something else went awry." Lifting a tower initially without the turbine is

FIGURE 6.16. Grip Hoist. Ian's son, Zander, demonstrating the use of a grip hoist. This device can be used to raise and lower tilt-up towers. Credit: Ian Woofenden.

FIGURE 6.17. Tower and Gin Pole. This photo shows how we raised the first section of our tilt-up tower. We added one pipe at a time, raised it, then adjusted the guy cables. This was very slow, but we didn't want to make any mistakes. Credit: Dan Chiras.

also important because it helps train your workers. That way, everyone knows what they are doing by the time the turbine and tower are lifted as one. Also be sure all workers and onlookers, including family pets, are out of harm's way.

Pros and Cons of a Guyed Tilt-up Tower

"My advice: if you have space for a tilt-up tower, use one!" advises Ian in "Wind Generator Tower Basics" in *Home Power* magazine. "You will never have to climb your tower (in fact, you won't be able to—the pipe or tubes are not climbable). All maintenance will be done on *terra firma*." When a wind turbine with a 10-foot diameter blade on an 120-foot tower needs to be inspected or serviced, you can lower the turbine and tower in less than half an hour, and have it back up and running in the same time—once you've completed the inspection or needed repairs. Taller towers and those supporting much larger turbines, a nearly 24-foot diameter Xzeres 442SR, for example, may take a little longer to raise and lower.

Although they're ideal for those who cringe at the idea of climbing a 80-to-120+-foot tower, tilt-up guyed towers do have a few drawbacks. For one, they have the largest footprint of all towers (Figure 6.9). This results from several factors. First, tilt-up towers require four guy cables that anchor in the ground some distance from the tower. The larger footprint is also due to the need for room to lay the tower down. A considerable amount of room is also needed to operate a lift vehicle. As a general rule, the guy radius is about 35 to 80 percent of the tower height. The gin pole is typically 75 to 100 percent of the guy radius. The lay-down zone is equal to the length of the tower, and the zone in which the lift vehicle operates is three times the height of the tower. For example, suppose you were installing a 120-foot guyed tilt-up tower. In this installation, the anchors would be placed 42 to 96 feet from the base of the tower, depending on the manufacturer's recommendations. The lay-down zone would be 120 feet. The vehicle access zone would be about three to five times the height of the tower, or as much as 480 feet. (That's another reason for using an electric winch.) This entire area would need to be free of trees and bushes so guy cables won't get caught in them as the tower is raised and lowered. The driving lane must also be free of obstructions.

Raising and lowering a tower isn't a job you should do alone. You'll need a couple of helpers to ensure that everything runs smoothly—for example, that the cables don't get tangled. And you'll need a truck, tractor, or some other lifting device.

Like all towers, things can go wrong with guyed tilt-up towers. They can get dropped while raising or lowering them and the wind turbine can be destroyed if it strikes the ground. Tow vehicles can slip. Accidents can also occur if the anchors are not correctly positioned or if the guy cables get too tight while lowering or raising the tower.

Another disadvantage of tilt-up towers is that, except for the lattice towers sold by Bergey, they are not climbable. This is a decided disadvantage if you are willing and able to climb and want to check out something simple on your turbine. Rather than climb the tower to repair it, you'll need to lower it and then raise it.

The top two sections of tall tilt-up towers are also pretty unstable during the first 5% to 10% of the descent. To stabilize them, and prevent towers from collapsing, a heavy-duty rope can be attached to the bottom of the first section from the top at the coupling. A cable is often attached to the bottom of the second section from the top. It is attached to a hand-operated winch. When the tower is being laid down, one worker applies tension to the rope while a second one applies tension to the cable. This keeps the first 40-feet of tower more stable and increases your chances of a successful tower lowering. The same process is repeated during the last 5% to 10% of the ascent.

Another disadvantage of tilt-up towers is that they require a lot of concrete and foundations. Steel rebar cages must be made and placed in the holes excavated for the anchor (Figure 6.18). Care must be taken when installing the right and left tower anchors to be sure they are level. These two anchors and their associated guy cables act as a hinge that allows one to raise and lower the tower. If the anchors are not at the same height, you could run into trouble raising and lowering a guyed tilt-up tower.

Even though Mick Sagrillo designed the first two tilt-up towers in the residential wind market many years ago, he prefers to climb towers rather than tilt them up and down. "Climbing puts you into a mental perspective where you are constantly on alert as you ascend or descend the tower, or work on the wind generator. You simply do not take risks at heights," notes Mick. Ground crews for a tilt-up tower operation, on the other hand, feel safe at ground level, and Mick has frequently seen their attention wander. This can be especially dangerous with inexperienced crews. "Just because you are on the ground doesn't mean you are safe," Mick warns. Should a distracted crew member lose control of the tower or should a problem occur with the tower, the consequences can be serious. The acceleration of a falling object—32 ft/sec² due to gravity—can result in a disaster *before* most people are aware that there is a problem. Anyone standing where a guy cable is headed on a tower that is no longer under the control of the winch or truck used for raising

or lowering a tower…well…use your imagination. Think "egg slicer," and you'll understand the possibilities. After having installed numerous tilt-up towers and conducted multiple tilt-up tower workshops, Mick has come to the conclusion that the presumed safety advantage of a tilt-up tower is a myth. In fact, more people have had close calls with tilt-up towers than climbable towers. And one final point about tilt-up towers: although they offer key advantages, you'll never be able to enjoy the view from the top of your tower.

Guyless Tilt-up Towers

In recent years, several US companies have introduced short, heavy-duty freestanding towers that can be lowered by hand, allowing inspection, maintenance, and repair at ground level. These are described by the manufacturers as either hinged, tilt-up, or tilt-down towers. In fact, search for images of tilt-down towers on the internet and you'll find an amazing number of homemade and manufactured

FIGURE 6.18. Making Anchors. (*a*) These rebar cages can be wired or welded together and then (*b*) placed in the holes excavated for foundations and anchors. (c) Concrete is then poured into the hole and allowed to set prior to tower installation. Notice the galvanized steel anchor (auger) embedded in the concrete. (*d*) The guy cables are attached to the anchor. Credit: Dan Chiras.

towers. One thing you will notice is that most of them are pretty short—well under 40 feet.

Tilt-down towers consist of two sections, joined by a hinge. One part remains erect; the upper portion pivots on the hinge, allowing it to be raised or lowered.

As noted in previous chapters, for best performance, most wind machines should be installed on 80- to 120+-foot towers, which raise machines into the strongest, smoothest winds. While a case could be made for installing short freestanding towers in flat terrain (for example, grasslands) or along lake or ocean shorelines where obstructions to wind flow are minimal, even in perfectly flat terrain, towers should be at least 80 feet high. I'm aware of only one guyless tilt-down tower. It's manufactured by a Texas company, Western Towers. If you are interested, be sure that it is sturdy enough to support the wind turbine you are considering.

Homebrew Towers

For those among us who like to do things themselves, the thought of making our own tower might seem appealing. There are some ingenious designs out there, and I have great respect for independent-minded souls and applaud the creative efforts of do-it-yourselfers. When it comes to making your own tower, however, I urge extreme caution. "This is no place for lightweight construction or engineering guesswork," Ian notes in his article on towers in *Home Power* magazine. "If you are going to build your own tower, do careful research. Look at engineered towers and get a sense of the designs, as well as the size and quality of hardware used."

"When in doubt," he continues, "overbuild. Better yet, stick with engineered towers that are professionally designed for the job. To obtain permits, you may need an engineer's stamp on your plans anyway."

Tower Kits

If all this discussion of towers seems daunting, don't be dismayed. If you are hiring a professional to install your wind system, he or she should pick out and install a tower that meets your needs. If you are installing your own wind turbine, note that several turbine manufacturers sell tower kits designed and engineered for their wind machines, although some may only be available to certified installers. Beware of many short-tower kits, however. Go straight to reputable manufacturers like Bergey and Xzeres that sell tall-tower kits for their turbines. When buying a tower made from pipe, remember: steel pipe is heavy, so it is expensive to ship long distances. As a result, most kits include all of the fittings and cables but not the pipe. Purchase the steel pipe locally.

FIGURE 6.19. Adaptor. This stub tower for my Skystream 3.7 fits between the turbine and the pipe of a tower. Don't forget to order a stub tower or adaptor when ordering a tower from a manufacturer. Credit: Dan Chiras.

When shopping, bear in mind that wind turbines often require adaptors, commonly referred to as stub towers, to fit onto commercially available towers. Stub towers consist of a short piece of pipe, with a flange that bolts onto the base of the turbine (Figure 6.19). Be sure to obtain an adaptor for your turbine. For best results, buy a hot-dipped galvanized adaptor. They last much longer than ungalvanized steel stub towers.

Remember, too, when installing a tower with guy cables the uppermost attachment point of the guy cables should be well below the rotor.

When buying a tower, be sure you purchase a model that's approved by the wind turbine manufacturer. Not doing so may negate the warranty! When shopping for a wind generator tower kit, also be aware that some kits may not include anchors, because the type of anchor needed for a tower varies from one location to the next, depending on the soil type, discussed shortly. In most cases, anchor foundations are constructed on site using concrete and rebar.

While I am on the subject of tower kits, I want to warn readers against buying kits to mount towers and wind turbines on roofs—or even against walls—of buildings. Such installations are an unequivocally bad idea. Why?

First of all, roof-mounted turbines in most locations are too close to the ground, where wind speeds are slower. Lower wind speeds mean lower output. A 2007 study of roof-mounted wind turbines by Encraft, Ltd., a British consulting firm, showed that wind turbines mounted on homes and apartment buildings produced much less electricity than predicted by computer models. On residential structures, annual wind speeds were only about 5 to 7 miles per hour, well below the models' predicted wind speeds of 10 to 12 miles per hour. (The models obviously didn't account for ground drag and turbulence).

Kilowatt-hour meters were also installed to monitor the output of the microturbines and small turbines in this study. The researchers compared energy generated by the urban turbines to the manufacturers' performance predictions. The researchers found that on average, wind turbines mounted on roofs and short towers exported less than 0.5 kilowatt-hours a day. Some of the turbines generated less power than the inverters consumed and were therefore negative energy producers at their low wind speed sites. That is, these systems consumed more electricity than they generated.

The second reason rooftop installations are not a good idea is that buildings and associated ground clutter like trees create turbulence. Turbulence reduces the quality and quantity of wind. Lower wind speeds and turbulent winds reduce energy production. Therefore, roof-mounted wind turbines will produce significantly less electricity than turbines mounted on tall towers. More turbulence also means more wear and tear on the turbine, more costly replacements, and shorter turbine life.

Third, buildings are rarely designed and engineered to support the load and handle the vibrations produced by a wind turbine. These vibrations can cause structural damage to buildings. They may cause screws holding panels and inverters to come loose, causing damage. They may cause framing members to detach as well. Mick once saw a garage with the rafter nails vibrated out by a roof-mounted wind turbine.

Fourth, noise conducted into the building can be disturbing if the building is occupied. I installed my very first wind turbine, a microturbine made by Southwest Wind Power, on a short tower (first mistake) against a retaining wall (second mistake) that abutted the house. At night, when the winds blew fiercely, I felt like I was living inside an acoustic guitar. The turbine growled in high winds and filled the house with an annoying sound.

So, as nifty as this idea may seem, stay away from rooftop towers. Even in unoccupied buildings, the problems with roof mounts greatly outweigh the benefits of this approach.

Tower Anchors and Bases

When installing a tower, you'll need to install a strong base. If the tower is guyed, you'll also need to install anchors for the guy cables. All reputable manufacturers provide well-engineered plans for suitable foundations. Follow the plans very carefully. Don't cut corners to save money or time. Use common sense. Take a wind workshop or two *before* installing your wind machine, or hire a professional to do the job.

Tower Base

Virtually all towers are mounted on concrete pads reinforced with rebar (Figure 6.20). The depth required for a foundation, known as the *critical depth*, is the depth that prevents a foundation (and anchors) from being pried out of the ground by the force of the wind or jostled by freeze-thaw cycles. The critical depth of a foundation depends on many factors, such as the type and the height of the tower, wind speed, depth of the frost line, and soil characteristics.

As noted earlier, freestanding towers require deeper and more robust foundations than guyed towers. The taller the tower, the stronger the foundation. The stronger the winds, the deeper and more robust the foundation. The deeper the frost line, the deeper the foundation. Foundations not installed below the frost line may be heaved out of the ground as the soil moisture freezes and expands. To

FIGURE 6.20. Concrete Base. Virtually all small wind turbine towers rest on a solid concrete base. This photo shows three concrete piers for the three legs of a freestanding lattice tower. Credit: Dan Chiras.

determine the frost line in your area, check with your local building department or a knowledgeable builder in your area.

Soil characteristics also affect foundation design and construction. Some soils hold tower foundations in place better than others. Sandy soils, for instance, have less holding power than clay-rich and heavier soils, so tower foundations need to be much deeper than those in heavier clay-rich or rocky soils. For advice on critical tower depth, contact the turbine and tower manufacturers.

As noted earlier, freestanding towers require the most concrete and steel. In a lattice tower, each leg of the tower is attached to a steel anchor that's embedded in a steel-reinforced concrete pier (Figure 6.20). The base of the pier is often wider than the column above it, creating a footing that distributes the weight of the tower. Alternatively, piers may be attached to an underlying concrete foundation that joins all three piers together. The pad acts like a footing and provides even greater resistance to the wind.

Freestanding monopole towers require a single deep, robust steel-reinforced pad located well below the frost line. Foundations for guyed and tilt-up towers require much less concrete and rebar, for reasons explained earlier. The most common concrete foundation for freestanding monopole towers consists of a single square or round concrete pad. The thickness of the pad varies from one site to the next, depending on climate (depth of the frost line), soil strength, wind speed, and tower height. Thin pads are used in warm climates. Deep, thick pads or other types of foundation are used in colder climates.

Engineered plans are typically provided in installation manuals that come with turbines. Even so, when installing a wind generator tower, be sure to consult a qualified structural engineer who specializes in towers and who understands local soil conditions to develop a tower foundation that's affordable yet sufficiently strong.

Concrete pads for tilt-up and guyed towers require hardware to attach the tower. To secure a guyed pole tower, for instance, installers embed anchor studs, 26-inch-long bolts threaded on both ends, in the pad. They are bolted into a one-half-inch steel anchor plate, shown in Figure 6.21, which is embedded in the concrete foundation, and attached to a flange plate on the base of the tower.

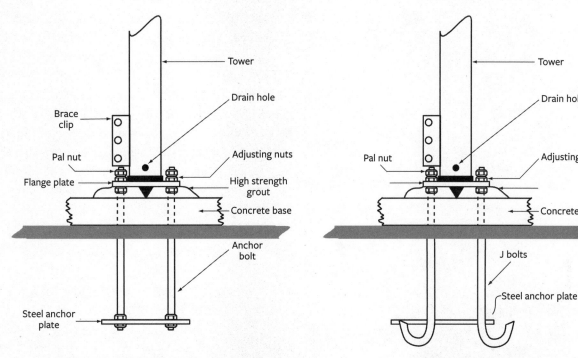

FIGURE 6.21. Tower Attachment Details. This tower is attached by anchor bolts. The anchor bolts connect the flange plate on the base of the tower to the concrete foundation. Credit: Illustration by Anil Rao and courtesy of Bergey WindPower.

FIGURE 6.22. J-Bolts. In this tower, J-bolts anchored in the concrete pad are attached to a flange plate at the base of the tower, thus securing the tower to the deep steel-reinforced concrete anchor. Credit: Illustration by Anil Rao and courtesy of Bergey WindPower.

Other foundations use longer J-bolt anchors embedded in the concrete pad (Figure 6.22). Be sure to follow the manufacturer's recommendation.

For maximum strength, many professionals recommend 28 days of curing prior to installation of the tower. Whatever you do, do not place a tower on a concrete pad before this time! Inadequately cured concrete weakens the strength of the concrete. So don't rush it! Your tower and wind machine rely on a sturdy, well-cured foundation (Figure 6.23). Remember, too, that cold weather slows the curing process. More time may be required in such conditions.

Anchors

In addition to a concrete base, fixed, guyed towers and guyed tilt-up towers also require anchors to attach the guy cables. Anchor options for wind generator towers are many and varied. For most wind turbines, concrete anchors are the best choice. A concrete anchor is created by digging a deep hole, installing rebar, and then

FIGURE 6.23. Tower Base at TEI. (*a*) This newly poured tower base at The Evergreen Institute was allowed to cure for a month prior to mounting the tower. Notice the J-bolts that were embedded into the concrete shortly after it was poured. (*b*) Be sure to space the bolts carefully so they match up with the bolt pattern of the tower. We built a template that was positioned over the newly poured concrete to ensure proper spacing. (*c*) Finished tower base, ready for tower installation. Credit: Dan Chiras.

pouring concrete into the hole, illustrated in Figure 6.18. An auger anchor or angle steel is installed prior to the concrete pour. It is angled toward the tower as per manufacturer instructions (Figure 6.24). Guy cables are attached to these structures (Figure 6.25). Once the concrete has cured, the hole is filled in. Be sure to pour anchors below the frost line and follow the advice of the tower manufacturer or supplier.

Augers embedded in the ground are additional options for tower anchors, but are typically only

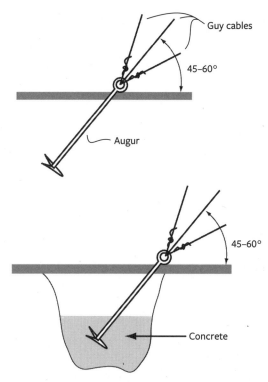

FIGURE 6.24. Auger. Augers can be embedded in concrete (*bottom*) to form a bombproof anchor for guy cables on guyed tilt-up and fixed, guyed towers. In some cases, augers can be screwed directly into the ground, securing the guy cables of a wind turbine (*top*). The latter create an economical anchor, but are of limited use as explained in the text. Credit: Anil Rao.

FIGURE 6.25. Guy Cables. Guy cables attach to the concrete anchor via a metal plate embedded in the anchor. Turnbuckles attach to the plate and the guy cables allow them to be properly tensioned. Credit: Dan Chiras.

suitable for very small wind turbines—the microturbines installed on short towers. (As I've stated many times, I frown on the use of short towers.)

Augers are manually screwed into the ground using a piece of pipe or a steel rod, known as a *cheater bar*, which makes it easier to turn the auger. (They provide leverage that reduces the force you must apply.) Augers should be screwed in at an angle tilted toward the tower as specified by the manufacturer.

Longer, heavy-duty augers, measuring four feet or more are appropriate for short wind generator towers (under 60 feet). They may be available through turbine and tower manufacturers or through utility supply houses.

Augers are easy to install, fairly inexpensive, and can also be removed if you make a mistake in placement. Augers are suitable for clay or loamy soil, but can be difficult to install in rocky soil, and they do not hold in sandy or gravely soil.

Another option for short towers is the duckbill. A duckbill is a small metal device that's attached to the guy cable and driven at an angle into the ground using a steel rod (Figure 6.26). Once it's been driven to the desired depth, the installer tugs on the duckbill. This turns it perpendicular to the surface. Like a deadman, it holds the cable securely in the ground.

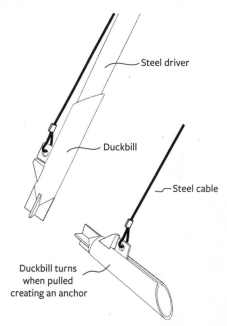

Steel driver

Duckbill

Steel cable

Duckbill turns when pulled creating an anchor

FIGURE 6.26. Duckbill Installation. A duckbill is driven into the ground using a steel rod and hammer (*top figure*). When it's deep enough, the installer tugs on the cable, which causes the duckbill to turn, locking the cable in the ground (*bottom figure*). Credit: Anil Rao.

Duckbills are very inexpensive and easy to install. However, they are only suitable for small turbines on short towers in rocky soils. They're a nightmare to remove, and the guy cables run underground, where they can rust.

For sites with lots of large rock or on solid bedrock, expansion bolts or large eyebolts can be cemented with epoxy into the rock. Expansion bolts can be used in harder rock such as granite and for softer rocks, such as the harder grades of sandstone, although longer bolts are necessary.

To insert an eyebolt, a hole must first be drilled into the rock. The bolt is then placed into the hole with an attachment hanger to connect to the guy cable. Epoxy is injected around the bolt. I recommend that you consult the tower manufacturer or a company familiar with rock anchors and rock structure before you install a wind generator tower using anchor bolts.

As you can see, there are a variety of anchor options available for wind generator towers; however, most of them are unsuitable for taller towers—80 to 120+ feet. For tall towers (the kind you should be using to install your wind machine), steel-reinforced concrete anchors are your best bet. They're stronger, long-lasting, and easy to construct.

Safety First

Wind energy is a great source of electricity. Unlike PV modules in a solar-electric system that provide years of maintenance-free service, a wind turbine requires routine inspection and maintenance and occasional repair. This, in turn, requires periodic ascent of the tower or lowering and raising the tower. Both activities pose some risk.

Installing a tower is also a risky business—even more so if it is your first experience. Safety of workers and onlookers should be an installer's primary concern during an installation. Moreover, nearby equipment and structures could be damaged if a tower collapses or topples during installation. And, lest we forget, wind turbines themselves can also be damaged if a tower collapses or falls to the ground during installation. At the very least, the blades are probably going to break in a fall. The generator could also be damaged. Your risks are reduced if you ensure that all parties remain aware of the dangers and observe all of the safety rules.

With this in mind, I offer some safety tips for tower installation:

1. Individuals not directly involved in raising the tower should stay clear of the work area. The job supervisor needs to monitor where people are standing and move people out of harm's way if they get too close.

2. All individuals on or in the vicinity of the tower should wear OSHA-approved hard hats. (Hard hats are meant to protect your head from bumping against something, not to protect your head from something dropped from a tower.)

3. Tower work should be performed by—or should be under the supervision of—a well-trained installer. This is tricky work; it requires knowledge and experience.

4. Towers should never be constructed near utility lines. If any portion of the tower or equipment comes into contact with them, serious injury or death may result.

5. People working on a raised tower should be roped in at all times via a safety harness and lanyard. Workers should always be connected to an anti-fall climbing cable and safety device when ascending or descending (discussed earlier).

6. Tools should not be carried in one's hands when climbing a tower. Use a hoistable tool bucket with a lid (Figure 6.27).

7. Never stand or work directly below someone who is working on the tower.

8. Never work on the tower alone.

9. Never climb the tower unless the wind turbine is shut down and the alternator is secured with a brake or by a short circuit. The best location for the short-circuit switch or jumpers is in the tower-base disconnect switch box.

10. Never drop or throw anything from the tower. If you drop lines when working on a tower, warn the ground crew and be sure to receive visual or auditory feedback that they've heard your warning. Be sure they have cleared the area.

11. Stay off the tower during bad weather or impending bad weather, including thunderstorms and high winds. Stay off the tower after ice storms.

12. If you see bad weather approaching, climb down immediately and suspend activities until the storm has passed.

FIGURE 6.27. Canvas Bucket. Lightweight and easy to carry, this bucket is used to transport tools up a tower and hold tools while service personnel are working on a wind turbine. Never carry a plastic bucket, because the handle could come off. Credit: Dan Chiras.

Siting a Tower

As you know by now, to generate electricity economically and in sufficient quantity to meet your needs, you need to place a wind turbine in a good wind site. For optimum electrical output, a wind turbine must be mounted well above the ground, out of the way of ground clutter, in the strongest, smoothest winds. Careful siting and proper tower height combat the two biggest enemies of wind systems: ground drag and turbulence, discussed at length in Chapter 2. Let's begin with proper siting.

Proper Siting of a Wind Machine

When siting a wind generator—that is, determining a location that will allow the wind turbine to perform optimally—remember at all times that your main goal is to situate the tower out of the turbulence bubble created by ground clutter.

Wind site assessors begin this process by determining the prevailing wind direction at a site. Although winds blow in different directions at different times of the year, even within the same day, they'll typically blow from one or two directions predominantly over the course of the year. In many places in North America, for example, the winds come predominantly from the southwest in the spring, summer, and fall—thanks to the Coriolis effect described in Chapter 2. In some locations, they blow from the northwest in the winter. How do you determine the predominant wind direction?

One way is to ask long-time residents for advice. Especially helpful are farmers, who work outdoors and hence are familiar with wind patterns. Another, potentially more reliable way to determine predominant wind direction is to contact a nearby airport. They may be able to provide you with a *wind rose*, a graphical representation of wind direction like the one shown in Figure 6.28. In a wind rose, the length of the spokes around the circle is an indication of how frequently the wind blows from a particular direction. The longer the line, the greater the frequency. In the example shown in Figure 6.28, the winds blow predominantly from the southwest. A wind rose may also indicate the percentage of total wind energy from each direction, which is very helpful. You can also find data on wind direction at the NASA Surface Meteorology site discussed in Chapter 4.

In an open site, with little ground clutter, a wind turbine can be located almost anywhere—so long as

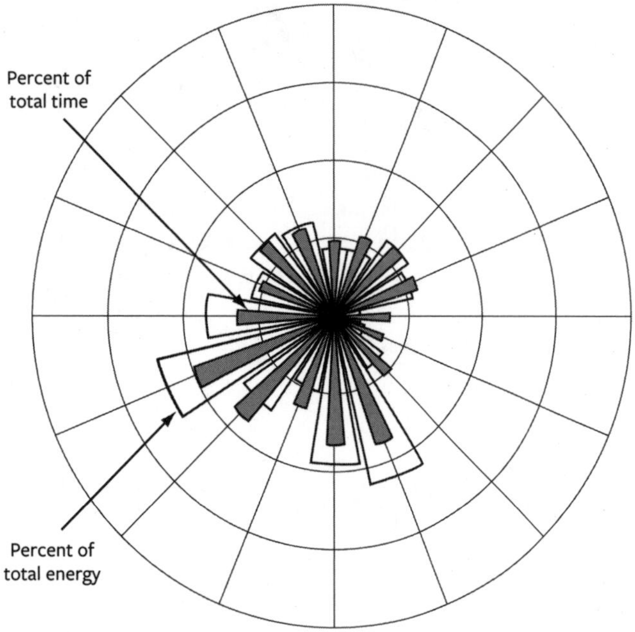

FIGURE 6.28. Wind Rose. This unique graph shows how often winds blow from various directions and the percent energy of the wind for various directions. The wider white bars represent the percent of total energy from different directions and the narrower, shaded bars illustrate the percent of total time from each of the 16 different direction sectors. Credit: Anil Rao.

the entire rotor is mounted 30 feet above the tallest obstacle within a 500-foot radius *and* you've taken into account future tree growth, if trees are the tallest objects. Unfortunately, very few of us live on ideal sites. There's almost always some major obstacle that makes siting difficult. What do you do in such instances?

Once you have determined the prevailing wind direction, look for a location for the tower upwind of major obstacles. Although winds will shift so that upwind becomes downwind at times, your wind turbine and tower will be situated so that it takes advantage of the strongest prevailing winds. You still need to obey the 30/500-foot rule. This is frequently most easily accomplished by installing your tower on one of the highest spots on your property.

When siting a wind turbine, it is also a good idea to minimize wire runs from the turbine to the controller and inverter to reduce line loss. As a rule, the higher the wind turbine's voltage, the farther it can be sited from the point of use. When installing a turbine, contact the manufacturer or an experienced installer to discuss wire runs. All manufacturers should supply a chart showing recommended wire size based on turbine type and length of wire run.

Tower Height Considerations

Once you've identified the best site, it's time to determine optimum tower height. As noted in previous chapters, for a wind system to perform optimally—to produce as much electrical energy at the lowest cost possible at a site—it needs to be on a tall tower. The 30/500 rule just mentioned stipulates that the *entire* rotor should be *at least* 30 feet above the tallest obstacle within a radius of 500 feet.

Consider an example. Suppose you were going to mount a wind machine on a farm with a 45-foot silo 250 feet upwind of the proposed tower site. Suppose the blades of the wind turbine are ten feet long. To determine the minimum acceptable tower height add the silo height (45 feet) to the blade length (10 feet) to the 30-foot height allowance. The minimum tower height is 85 feet. Since towers come in 10-foot sections, round up to at least 90 feet. Remember, however, that 85 feet is the minimum acceptable tower height. Because output increases on taller towers, I strongly recommend that you also analyze the cost and output of the same turbine on a 100- and also a 120-foot tower.

When calculating minimum tower height, don't forget to take tree growth into account. If the trees on your property will grow 20 feet in the next 20 to 30 years (the life expectancy of a wind system) add that to the tower height for the best long-term performance. Anything less will reduce the performance of the wind generator and reduce its life. For example, suppose you were going to install a wind turbine

How High?

Determining the height of a tower seems pretty straightforward until you have to do it. The first challenge you'll face is determining the actual height of nearby objects, such as trees or barns. How do wind site assessors determine the height of ground clutter?

One way, shown in Figure 6.29, is to place a stake (a metal fence post, for instance) next to the object you want to measure. On a sunny day, measure the height of the stake and then measure its shadow. Then measure the shadow of the object under question.

The height of the object can be determined by ratios using the equation: $H_1/H_2 = SL_1/SL_2$. H_1 is the unknown height and H_2 is the height of the fence post.

SL_1 is the length of the shadow of the object you are trying to measure; SL_2 is the length of the shadow of the fence post. To solve for H_1, you'll need to rearrange the equation: $H_1 = (SL_1)(H_2)/SL_2$. Note that the ground around both the tree and the fence post must be level for this method to be accurate.

Consider an example. Let's assume that the fence post is four feet high and the shadow it casts is two feet long. The shadow cast by the tree or building is 14 feet. How high is the tree? As illustrated, in Figure 6.29a, you begin by setting up ratios: $14/2 = x/4$, then solve for x: $x = (4 \times 14)/2 = 28$ feet. Another simple method is explained in Figure 6.29b.

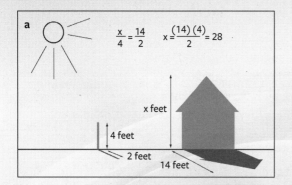

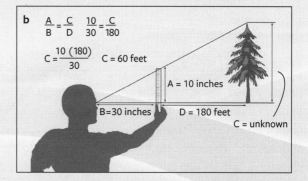

FIGURES 6.29. Measuring Height. (a) Driving a fence post or some other object of known length into the ground next to an object of unknown height and comparing the length of the shadows, allows one to calculate the height of an object. (b) Another method for determining height is shown here. In this method, you'll be solving for C, the height of the tree. Measure the distance from the tree (D). Measure the distance from your eye to the ruler in your hand. This is B. Measure the height of the object in inches on the ruler. This is A. Then set up a ratio equivalence as follows: A/B = C/D. The rest of the math is shown in the figure. Credit: Anil Rao.

within 500 feet of a 50-foot tree whose mature height is 75 feet. In this case, the tower should be at least 115 feet. That's 75 feet for the tree height added to the blade length (10 feet) plus the 30-foot rule. Since the tower you want in this example comes in 20-foot sections, round up to at least 120-feet. A turbine on a 140-foot tower would perform even better! (To learn two ways to estimate the height of trees and buildings, see the accompanying textbox.)

If your site is within a quarter of a mile from a forest or good-sized wood lot, the top of the nearby tree line is the height you want to exceed. Mount the wind turbine as just described, using the tree line as the height you must exceed. And remember that the trees will very likely grow.

To determine tree growth, you can check out the mature height of the trees in the area. If the trees on your property are 50 feet tall and top off at 65 feet, use this number. Anything less will reduce the performance and life expectancy of your system. You can find information on mature tree height in field guides on trees and on the internet.

While we're on the subject of tree growth, remember that tree growth varies in response to climate, especially precipitation. In wetter climates, trees grow more rapidly than in drier areas. Take this into consideration.

Determining tower height in heavily wooded areas can be tricky. If you are installing a wind turbine in an area with more than 50 percent tree cover, the effective ground level is two-thirds of the tree height. If trees are 60 feet high, for instance, the effective ground level, known as the *displacement height*, is $0.66 \times 60 = 40$ feet. A 100-foot equivalent tower would, therefore, need to be 140 feet high to take into account the trees.

In the literature on wind energy systems and in product literature, you will see statements contradicting my advice on turbine height. It is not unusual, for instance, to read that a wind machine should be placed at least 20 feet above obstructions within 250 feet.

For best performance, the acceptable numbers are 30 feet and 500 feet. Bear in mind, too, that this is the *minimum* acceptable tower height. Taller towers are generally better. Savvy wind energy installers generously exceed the basic rule of thumb, and see increased performance because of it. It usually costs very little to increase tower height by another 20 to 40 feet, and the return on this small investment is usually quite impressive. I don't know anyone who has installed a wind turbine who says, "I wish I'd bought a shorter tower." But I (and others in the industry) know lots of people who wish they had purchased a taller tower.

FIGURE 6.30. Wind Machine vs. Trees. This wind machine no longer produces any energy; the trees around it have grown taller than the turbine and successfully block virtually all wind. Credit: Mick Sagrillo.

Figure 6.30 shows a wind turbine that was installed in the early 1980s two miles from Mick Sagrillo's home in Wisconsin. At the time of the installation, the trees were well below the height of the house. The installers, Mick, and the owner, were greenhorns and didn't take tree growth into account. The result?

Today, the wind machine is hidden from the wind. The blades of the wind machine are so sheltered that they only occasionally spin, and the machine produces no electricity. It's become an expensive lawn ornament. So why doesn't the owner just raise the tower?

Unfortunately, it is hard to "grow" towers, unless you've planned and engineered for it in advance. Tower anchors and guy radiuses are only correct for a specific height. If you decide to go higher later, you usually will need to start over from scratch—which will be expensive.

Tall Tower Economics: Overcoming the Small-Turbines-on-Short-Tower Myth

When you talk to professional wind system installers, you may hear statements to the effect that it doesn't make sense to mount a smaller turbine, for example, one with a seven- or eight-foot diameter rotor, on a tall tower.

Wind energy experts strenuously object to such statements. Tower height should be determined by the height of obstructions in the area, not the size of the wind turbine or the towers a manufacturer or dealer sells. A 50-foot tower slightly downwind from 65-foot-high treeline isn't going to produce much electricity. Moreover, the turbine will produce even less electricity as the trees grow over the 20-to-30-year life of the wind system. Remember: energy output and the economics of the wind system are both proportional to V^3 (the cube of the wind speed).

Although it is sometimes hard to justify a tall tower for a small turbine, that doesn't mean that the right decision is a short tower. The right decision is to invest enough in your tower to make the most of your turbine's potential—or choose another renewable energy system.

If you are thinking about installing a smaller wind generator, but are nervous about the cost of a taller tower, calculate how much more the tower will cost and how much more electricity the turbine will produce on a taller tower. In my experience, installing a taller tower always results in the production of more electricity. Even though it will always cost more money, the important question to ask is whether the increased tower height is justified economically by the increase in electrical production. In most cases, it is.

Aircraft Safety and the FAA

Another factor to consider when installing a wind turbine is its impact on aviation and the possible need to file for a permit from the Federal Aviation Administration (FAA).

Fortunately, the FAA requires a permit application under only two conditions. The first is when the height of the turbine and tower is over 200 feet, which is extremely rare for small wind turbines. The 200-foot requirement on a wind turbine and tower refers to the total height of the turbine and tower with a blade in an upright position. To remain below the 200-foot limit, then, the tallest tower a wind turbine with a 30-foot rotor diameter (15-foot blades) could be mounted on would be 185-foot (185-foot tower plus 15-foot blade = 200 feet). Most home-sized wind towers do not exceed 120 feet.

The tallest towers offered by manufacturers and dealers as "off the shelf" items are in the 100- to 120-foot range, well below the 200-foot mark that could trigger a review by the FAA. In instances where tower heights exceed the 200-foot mark, the FAA may require top-of-tower lighting to warn pilots.

The second situation that would require an FAA review and permit is based on the proximity of a wind generator tower to an airport—specifically a "public use" or military airport. Permit applications are required if you are installing a turbine within 3.79 miles of an active public use or military airport if the runway is longer than 3,200 feet. If the runway is less than 3,200 feet long, you'll need to file an application if a turbine is being installed within 1.9 miles of the runway. In either case, the FAA will determine the height of the tower you can install. They may also

FAA Applications and Zoning Hearings

The need for an FAA application sometimes arises during zoning hearings for small-scale wind systems in an attempt to delay the installation of a turbine and tower. However, this is a matter between the applicant and the Federal Aviation Administration. Zoning officials should not use the FAA application, if any, to delay the installation of a wind turbine and tower, or as an added hurdle to dissuade homeowners from installing wind turbines.

require top-of-tower lighting. Bear in mind that this requirement does not pertain to private landing strips with no public access, airfields not shown on FAA maps, or landing strips that are not in active use.

An FAA application takes at least several months to process, so plan ahead. If you are in doubt about your distance from an airport, or the status of the airport, it is always best to submit an application. If you're hiring a professional installer, he or she can advise you on this matter. The application form is #7460 (Notice of Proposed Construction or Alteration). Instructions are available at the FAA website, faa.gov. Addresses of the regional FAA offices where applications should be mailed are listed on the instructions.

If your project is outside the FAA guidelines, do not submit an application. The FAA does not want to be bothered with the permitting of a small wind system.

Protecting Against Lightning

When installing a wind turbine, it is important—indeed, essential—to include lightning protection, specifically ground rods and lightning arrestors.

As air masses move over the surface of the Earth, a static charge builds up between the Earth and the atmosphere, essentially creating a giant capacitor that stores static charge. At some point the static electricity has to equalize. It does so between 2,000 and 3,000 times a minute somewhere on the planet. This is the phenomenon called lightning.

Although lightning is not necessarily attracted to tall metal objects such as a wind generator tower, bleeding off the static charge by installing ground rods will reduce the likelihood of a strike. Ground rods are 8-foot-long copper-coated steel rods. They are driven into the ground at the base of the wind turbine tower. For guyed towers, you'll need a ground rod at the base of the tower and at each anchor to which the guy wires are attached.

For a three-legged freestanding tower, you'll need three ground rods, one for each leg. For a fixed, guyed tower, you'll need four ground rods, one for the tower and one for each of the three guy anchors. For a guyed tilt-up tower, you'll need five ground rods; one for the tower and one for each of the four guy cables.

Grounding a structure's static charge does minimize lightning strikes, but it cannot guarantee that lightning will not strike a tower. Backup is needed in the form of lightning arrestors and surge arrestors.

Lightning arrestors are known in the trade as *silicon oxide varistors* or *SOVs*. An SOV is a spark gap arrestor. Inside the SOV is a device that looks like a spark plug. It is embedded in white sand. One end of the "spark plug" is attached to the wind turbine's wiring; the other end is connected to the ground rod at the base of the tower. A spark jumps the gap when a certain voltage is reached. In theory, then, if lightning strikes a tower and travels down the wiring toward the battery bank or inverter in your house, the SOV shunts that lightning strike to ground. Mick stresses "in theory" because, like ground rods, there's no guarantee that a lightning arrestor will work every time. Sometimes lightning doesn't follow human rules and goes where it wants to.

SOVs should be installed on the electrical wire running down the tower as well as the utility wiring for grid-tied systems to protect against lightning strikes in utility lines. In reality, lightning strikes are orders of magnitude more frequent on the utility grid than on properly grounded towers.

While protecting against direct lightning strikes is important, nearby strikes pose the gravest danger to a wind system. Lightning is an electrical current in the atmosphere. This current creates a magnetic pulse that radiates out from the strike. If this magnetic pulse crosses a conductor like a wind generator tower or buried wires, an electrical current will be created (induced) in that conductor. This current can travel along a wire and fry sensitive electronic components of a wind system like a controller or inverter.

To protect against this problem, installers install *metal oxide varistors*, or *MOVs*, on the tower wiring—that is, the electric wires that run from the turbine to the base of the tower. (MOVs are used in computer surge arrestors to offer the very same protection to the sensitive electronics in our homes from surges that occur on utility or telephone lines.)

Will these measures guarantee that your turbine will never be struck by lightning? Maybe.

Lightning travels miles through the atmosphere, but the atmosphere is a poor conductor, so lighting can pretty much do as it pleases. Ground rods, SOVs, and

MOVs may help, but they aren't fail-safe. By installing these protective devices on a system, you have installed all of the precautionary devices offered by our best understanding of the science of lightning. You have done all that a prudent home-owner can do. If your system sustains damage, you should receive compensation from the insurance company.

Conclusion

Now that you understand wind turbines and towers, it is time to focus on some of the other key components of wind energy systems. In the next chapter, we will focus on inverters.

CHAPTER 7

Understanding Inverters

The inverter is a modern marvel of electronic wizardry and an indispensable component of virtually all renewable energy systems, wind included. Like batteries in off-grid systems, the inverter works long hours, providing us with electricity. As you recall, its main function is to convert DC electricity produced by the controller into AC electricity, the type used in our homes and businesses. In battery-based systems, inverters contain circuitry to perform a number of additional useful functions.

In this chapter, we'll examine the role inverters play in wind energy systems and then take a peek inside this remarkable device to see how it operates. We'll discuss all three types of inverters and discuss the features you should look for when purchasing one. But the first question to answer is this: Do you need an inverter?

Do You Need an Inverter?

Although this may seem like a ridiculous question, it's not. Some wind systems are designed to operate solely on DC power and, as a result, don't require inverters. Included in this category are small systems, like those in cabins, cottages, and sailboats, that power a few DC circuits. It also includes direct water-pumping systems that produce DC electricity to power a DC water pump.

All other renewable energy systems require an inverter. The type of inverter one needs depends on the type of system. Grid-connected systems require a utility-compatible inverter. Off-grid systems require inverters designed to operate with batteries. Grid-connected systems with battery backup require an inverter that's grid- *and* battery-compatible.

How Does an Inverter Create AC Electricity?

To help readers understand how inverters work, let's start with inverters installed in off-grid systems, known as *battery-based inverters*.

Battery-based inverters perform numerous functions. In off-grid systems, electricity from the wind turbine is fed into a charge controller. It then flows into the batteries. When electricity is needed, DC electricity stored in the batteries flows to the inverter. The inverter converts the DC electricity into AC electricity, typically 240 volts.

Batteries are wired so they receive, store, and supply 12-, 24-, or 48-volt electricity. The most common wiring configurations are 24 and 48 volts. In a 24-volt battery-based system, the inverter converts 24-volt DC from the battery bank to 120-volt or 240-volt AC electricity—either 50 or 60 Hz, depending on the region of the world you're in. (In North America, AC is 60 Hz; in Europe and Asia, it's 50 Hz.)

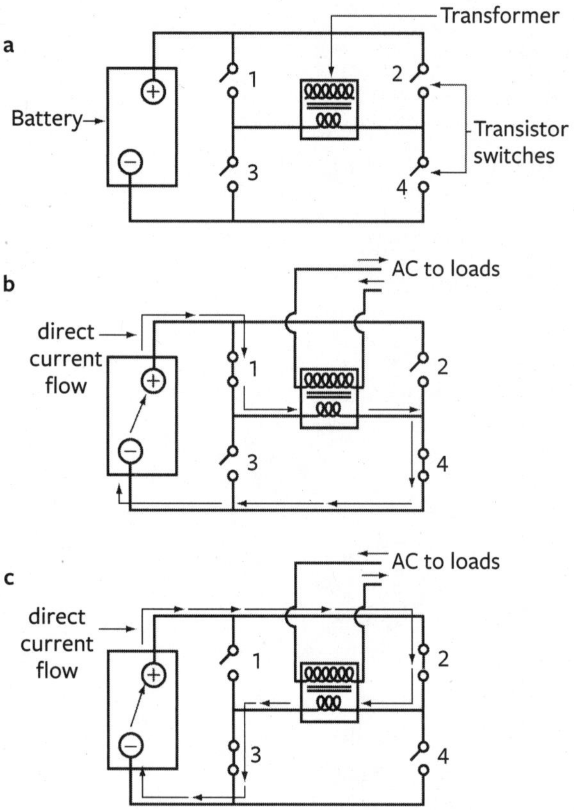

The DC-to-AC conversion and the increase in voltage can be performed by two separate, but integrated, components in an inverter. The conversion of DC to AC occurs in an electronic circuit referred to as an *H-bridge*, for reasons that will be clear shortly. (It is also known as a *high-power oscillator* or *power bridge*). The second process, the increase in voltage, is performed by a transformer inside the inverter. As you shall soon see, however, many modern inverters have done away with the transformer. The new inverters are known as *transformerless inverters*.

The Conversion of DC to AC

As shown in Figure 7.1a, the H-bridge consists of two vertical legs connected by a horizontal section, forming the letter H. Each leg contains two transistor switches that control the flow of electricity through the H-bridge.

To see how the H-bridge works, take a look at Figure 7.1b. As illustrated, DC electricity from the battery bank flows into the inverter. With switches 1 and 4 closed, low-voltage DC electricity flows from the battery into the vertical leg on the left side of the H-bridge then flows through the horizontal portion of the H-bridge

FIGURE 7.1. Diagram of Inner Workings of an Inverter: *a* (no electricity), *b*, and *c*. Credit: Anil Rao.

and then down the lower part of the right leg of the H-bridge. It then returns to the battery.

An electronic timer then flips the switches, opening 1 and 4 and closing 2 and 3. As shown in Figure 7.1c, electricity now flows from the battery into the vertical leg of the H-bridge on the right side of the diagram through switch 2. It then flows through the horizontal portion of the H-bridge (in the opposite direction) and then down the lower part of the vertical leg on the left side through switch 3. From there, it flows back to the battery.

Opening and closing the switches very rapidly allows DC electricity to flow first in one direction and then the other through the horizontal portion of the H-bridge and transformer. This alternating flow is AC electricity. A small controller (not included in the figures) regulates the opening and closing of the switches in the H-bridge.

This clever device creates 12-, 24-, or 48-volt AC electricity. As you can see by studying the diagrams, however, it is contained within the inverter. So, how does the inverter transfer AC electricity to the main service panel? And how is the voltage increased?

Increasing the Voltage

As illustrated in Figure 7.1, the horizontal portion of the H-bridge forms loops (for simplicity, I've only included three loops). These loops are referred to as a *winding*. (A winding consists of single wire, wrapped many times around a core.) This winding lies next to a secondary winding. Together, the two windings form a *transformer*.

A transformer is a rather simple electrical device that can increase or decrease the voltage of AC electricity. The transformer in a 24-volt inverter increases the 24-volt alternating current flowing through the H-bridge to 120-volt AC. The label "AC to loads" in Figure 7.1b and c indicates how the 120-volt AC electricity produced by the inverter exits. Those interested in learning how a transformer works should read the textbox, "How Transformers Work."

Square Wave, Modified Square Wave, and Sine Wave Electricity

An inverter, such as the one we've been examining, produces alternating current electricity. However, as shown in Figure 7.4a, it is very choppy. This graph is a plot of the voltage of the electricity over time. To understand what this graph represents, let's step back a second and review what we know about alternating current electricity. Electrons flow back and forth in a wire carrying alternating current at a rate of 60 cycles per second. If you measured the voltage in a wire carrying AC electricity

How Transformers Work

Transformers are used to increase (step-up) and decrease (step-down) voltage in the electrical grid and electrical devices. If you're tied to the electrical grid, chances are there's a transformer on a pole outside your home (Figure 7.2). This is known as a *step-down transformer* because it decreases the 7,500–24,000-volt electricity in the electric line running by your home (the voltage depends on your location) to 240 volts, the voltage used in homes. This transformer can also operate in reverse. That is, it can boost 240-volt AC backfed onto a grid from an inverter to match grid voltage. (That's one reason why inverters are designed

FIGURE 7.2. Step-down Transformer. This pole-mounted utility transformer reduces the voltage of incoming electricity, stepping it down from 7,500–22,000 volts (depending on the utility) to standard household voltage 240. This transformer steps up the voltage of electricity flowing from a wind system to the grid, as explained in the text. Credit: Dan Chiras.

to shut down when the grid goes down. This prevents high-voltage electricity from being backfed onto a dead grid.)

Step-up transformers, like the one found in an inverter, do just the opposite. That is, they increase the voltage.

Both step-up and step-down transformers use *electromagnetic induction*.

Electromagnetic induction is the production of an electric current in a wire as the wire moves through a magnetic field. This amazing phenomenon is the basis of virtually all electrical production technologies from coal- and natural-gas-fired power plants to wind turbines. For example, in the wind turbine shown in Figure 7.3, magnets are mounted in a rotating can, known as a *rotor*. It spins when the blades of the turbine spin. The magnets of a wind turbine rotate around a set of windings, numerous coils of copper wire. They are the stationary portion of a wind turbine generator, known as the *stator*. The rotation of magnets past the stationary (stator) windings produces electric current in the windings via electromagnetic induction. Cutting through the magnetic field literally pushes electrons in the copper wire, creating a current.

Magnetic fields are also produced as electricity flows through conductors. As you may recall, in an AC circuit, electrons flow back and forth, or cycle. As a result, in conductors carrying alternating current the magnetic field forms as the electrons race in one direction. When they stop, the magnetic field collapses. The electrons then reverse direction, and a new magnetic field forms. It collapses once again, though, when the electrons in the wire come to a stop at the end of the second half of the cycle. This alternating magnetic

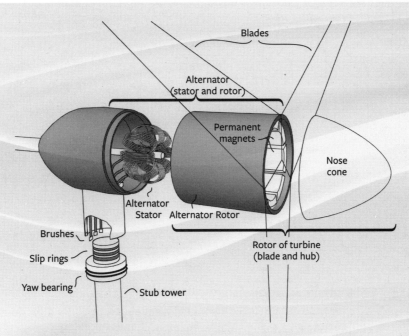

FIGURE 7.3. Diagram of a Wind Turbine Showing Alternator. Small wind turbines, like the one shown here, contain a set of magnets that rotate the blades of the turbine in the wind. These magnets rotate around several copper wire coils—the windings of the generator. As the magnetic fields pass by the windings, they generate AC electricity. Credit: Anil Rao.

field—that is, expanding and collapsing magnetic field—induces an alternating current in a nearby wire, or in the case of a transformer, the secondary winding.

As already noted, transformers consist of two *windings*, one that's fed electricity, and another that produces electricity at a higher or lower voltage. AC electricity flowing through the first coil, known as the *primary winding* creates an oscillating (expanding and collapsing) magnetic field. This oscillating magnetic field induces an electrical current in the secondary winding. In other words, it has the same effect as moving a magnet back and forth past a wire. How does a transformer increase or decrease voltage?

The voltage in the secondary winding coil increases and decreases in proportion to the windings ratio, that is the number of turns in the secondary winding compared to the number in the primary winding. The greater the number of turns in the secondary winding, the higher the voltage. If the ratio of windings is 2:1 the voltage will double. If the ratio is 3:1 the voltage will triple.

If the ratio is reversed, and there are fewer windings in the secondary coil than the primary coil, the voltage drops accordingly. If there are half as many windings in the secondary winding as in the primary winding, the voltage decreases by half.

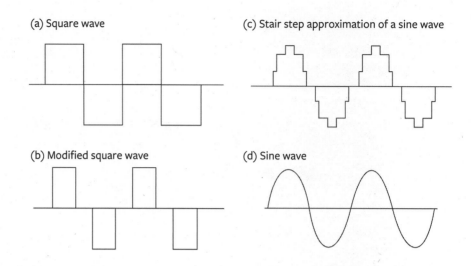

(a) Square wave

(b) Modified square wave

(c) Stair step approximation of a sine wave

(d) Sine wave

with an instrument that responds fast enough, you'd see that the current changes over time. It has to because the flow of electrons shifts direction—albeit very rapidly. When the electrons are flowing one way, say to the right, the voltage rapidly climbs from 0 volts to +120 volts. The electrons momentarily stop, and the voltage then drops back to 0. However, the flow of electrons quickly shift direction, the voltage rapidly rises again to –120 volts. The graphs of voltage in Figure 7.4 trace the change in voltage over time. However, you will note that the voltage goes from 0 to +120 volts, back to 0, then up to –120 volts. What do the positive and negative signs mean?

The positive and negative signs refer to the "polarity" of the voltage. More precisely, it is a human construct that says the current changes direction. The electricity is not charged differently, it just travels in different directions. In Figure 7.4a, the +120-volt plateau means the electricity is flowing to the right and the voltage is 120. (The horizontal line is 0 volts, voltage readings above that line are considered positive and readings below the line are negative.) The –120-volt simply signifies that the electrons are moving in the opposite direction for this brief moment.

The graph in Figure 7.4a indicates the voltage over an extremely brief period. The voltage shift from zero to +120 volts then back down to zero takes 1/120th of a second; the voltage shift from 0 to –120 takes another 1/120th of a second. Together, one complete cycle is 1/60th of a second, or 60 cycle per second electricity.

The basic H-bridge circuit found in some older inverters produces an almost perfect *square wave*. What that means is that the voltage jumps from 0 to +120 in-

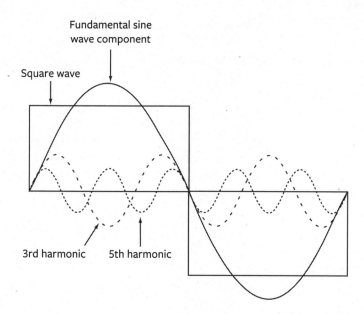

Fundamental sine
wave component

Square wave

3rd harmonic   5th harmonic

FIGURE 7.5. Harmonic Distortion. Square wave, and, to a lesser extent, modified square wave, electricity contains harmonics—waveforms at frequencies that are multiples of the desired frequency, as shown here. Motors, transformers, and many electronic devices do not perform well with such input. Credit: Anil Rao.

stantaneously, remains there for a short period, then instantly drops back to 0. Unfortunately, square wave electricity does not match the type of electricity produced by power plants. They produce *sine wave electricity*, shown in Figure 7.4d. Square wave electricity is not a very usable form of electricity. However, if this square wave is modified so that it pauses at zero volts briefly before reversing direction, the result is *modified square wave electricity*, shown in Figure 7.4b. Although doesn't appear to be much different, it is closer to a *sine wave*, the kind used in most homes. It can be used in off-grid homes, but has some serious shortcomings.

The problem with square wave and modified square wave electricity is that they contain *harmonics*. Harmonics are waveforms at frequencies that are multiples of the desired frequency. What that means is that there are electrons that cycle at different frequencies—besides the fundamental 60 cps.

Square wave electricity contains odd harmonics—that is, frequencies that are 3-times, 5-times, 7-times, etc. the fundamental frequency of the square wave (Figure 7.5). The output of a 60 Hz square wave or modified square wave inverter, for example, contains not only 60 cps, or 60 Hz, but also 180 Hz, 300 Hz, 420 Hz, etc. power. Motors, transformers, and many electronic devices perform poorly in such circumstances. They can even be damaged by this phenomenon called *harmonic distortion* (see sidebar, "What Is Total Harmonic Distortion?").

What Is Total Harmonic Distortion?

Total harmonic distortion (THD) is a percentage of harmonics in an inverter output—the frequencies we don't want—in relation to the frequency we do want. Electricity from a square wave inverter has a THD of about 48%. The output of a modified square wave inverter is somewhat lower than this, although you won't find a value on the inverter specification sheet. (Inverter manufacturers don't like to talk about the THD of their modified square wave inverters.)

In sharp contrast, the THD of the output of a sine wave inverter is 5% or less, about the same as the distortion in utility power. In some cases, the THD of a sine wave inverter is lower than that in utility power.

The amount of distortion on power lines is dependent on where you live and what your neighbors—and the utility—are doing. Harmonic distortion is produced by variable speed electric motors in factories and electric welders, arc furnaces, and computers. They all produce short bursts of noise or distortion in the line. In homes and small businesses, harmonic distortion is produced by fluorescent lights, computers, and some power cubes. Harmonic distortion also arises in electric utility systems from overloaded transformers and other technologies. A perfect sine wave has 0% THD.

Electronic filtering of the output of a square wave or modified square wave inverter can reduce these unwanted harmonics, but not eliminate them.

While modified square wave electricity works in many electrical devices like lightbulbs, toasters, microwave ovens, and blenders, most devices run less efficiently on this type of electricity. Because of this, virtually all inverters sold in North America and Europe produce a much purer form of electricity known in the trade as *sine wave electricity*. As just noted, it is nearly identical to the type of electricity delivered via the utility grid. If you are planning on connecting to the grid, you must install a sine wave inverter. How do sine wave inverters produce such clean power?

Sine wave inverters contain multiple H-bridges, each of which produces square wave electricity (Figure 7.6). However, the H-bridges are stacked on top of one another, that is, carefully spaced to produce a stair-step waveform that closely resembles a sine wave, as shown in Figure 7.4c. What that means is that each H-bridge creates a slightly different voltage. Their output is combined or stacked to form the stair-step waveform. Filtering this odd waveform produces a very clean sine wave, the kind of electricity our modern homes require (Figure 7.4d).

Transformerless Inverters

In recent years, many manufacturers have begun to transition to *transformerless inverters*. Unlike their predecessors, transformerless (TL) inverters rely on elec-

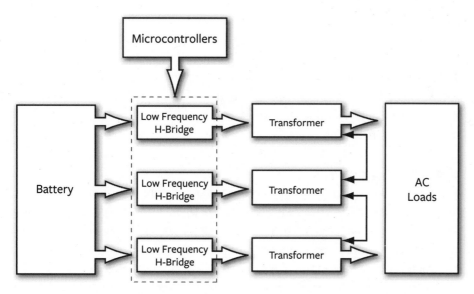

FIGURE 7.6. Multiple H-Bridges. Multiple H-bridges are found in transformer-based sine wave inverters. Each produces square wave electricity. Each H-bridge produces different voltages of AC electricity. The output of multiple H-bridges is combined to produce a stair-step approximation of sine wave electricity. It is filtered to produce pure sine wave electricity. Credit: Anil Rao.

tronic components that convert DC to AC. Like so many things, the conversion is controlled by a computer inside the inverter. The process requires several steps.

Part of the impetus for this shift was to produce smaller, lighter inverters that cost less to ship and are easier to install than their heavier counterparts. (From my experience, though, they're still pretty heavy.) The other impetus was to reduce costs and increase efficiency. According to SMA, which produced its first TL inverter in 2010, "transformerless inverters use electronic switching rather than mechanical switching, [so] the amount of heat and humidity they produce is much less than standard inverters."

TL inverters are designed for solar-electric systems, so those who are interested in installing hybrid wind and solar systems may find this information useful. If you are only interested in wind systems, you can skip this material. One important feature of TL inverters is that they are capable of being wired into two to four different arrays—for example, two to four solar-electric arrays that have different orientations and/or tilt angles.

Transformerless inverters require slightly different wiring than standard inverters, so be

Modified Square Wave or Modified Sine Wave?

A modified square wave inverter modifies—or cleans up—the square wave output from the H-bridge to reduce harmonics. Inverter manufacturers, however, usually label their modified square wave inverters as *modified sine wave inverters*. Why? It's marketing ploy. Customers know that a sine wave is better than a square wave, so an inverter that produces modified sine wave sounds better than one described as producing a modified square wave. In reality, they are the same.

sure to familiarize yourself with these requirements if you install a solar-electric system consisting of differently oriented arrays.

Types of Inverters

Inverters come in three basic types: those designed for grid-connected systems, those designed solely for off-grid systems, and those made for grid-connected systems with battery backup.

Grid-connected Inverters

Batteryless grid-connected wind-electric systems require inverters that operate in sync with the utility grid. Two types are available, string inverters and microinverters. String inverters designed for solar-electric systems can also be used in wind systems. Microinverters are designed only for solar-electric systems. Both types of inverter produce grid-compatible sine wave AC electricity that can be backfed onto the grid. Moreover, the electricity they produce is indistinguishable from grid power; often, it's even a bit cleaner.

Table 7.1. Wind System and Inverter Terminology

Common terminology used in this book (referring to systems and inverters)	UL 1741 (Refers to inverter)	NEC 690-2 (Referring to system)
Grid-connected	**Utility-interactive:** Operates in parallel with the utility grid.	**Interactive:** A solar photovoltaic system that operates in parallel with and may deliver power to an electrical production and distribution network.
Grid-connected with battery backup	**Multimode:** Operates as both or either stand-alone or utility interactive.	(Interactive, not separately defined)
Off-grid	**Stand-alone:** Operates independent of utility grid	**Stand-alone:** A solar photovoltaic system that supplies power independently of an electrical production and distribution network.
		Hybrid: A system composed of multiple power sources. These power sources may include photovoltaic, wind, micro hydro generators, engine-driven generators, and others, but do not include electrical production and distribution network systems. Energy storage systems, such as batteries, do not constitute a power source for the purpose of this definition.

As you learned in Chapter 5, grid-connected inverters, also known as *utility-tied inverters*, convert DC electricity from the controller in wind-electric systems into AC electricity (Figure 7.7). Electricity then flows from the inverter to the main service panel. It is then fed into active branch circuits, powering refrigerators, computers, stereo equipment, and the like. Surplus electricity, if any, is backfed onto the grid, running the electrical meter backward.

To produce sine wave electricity that matches the grid both in frequency and voltage, these inverters continuously monitor the voltage and frequency of the electricity on the utility lines. If an inverter detects a slight increase or decrease in voltage or frequency, it adjusts its output to conform. If, for example, the voltage drops from 243 volts to 242.7 volts or the frequency drops from 60 to 59.8 cps, the inverter automatically adjusts its output to conform to the grid. That way, electricity backfed from a wind or hybrid solar and wind system onto utility lines is identical to the electricity utilities are supplying to their customers.

Grid-compatible inverters are equipped with *anti-islanding protection*. Anti-islanding consists of hardware and software that automatically disconnects the inverter from the grid in case of loss of grid power. That's its primary purpose. All grid-connected inverters are programmed to shut down if the grid goes down until service is restored. As noted in Chapter 3, grid-tied inverters terminate the flow of electricity to the electric lines to protect utility workers from electrical shock, even electrocution (fatal shock).

Grid-tied inverters also shut down if there's an increase or decrease in the frequency or voltage of grid power outside the inverter's acceptable limits. These conditions are referred to as *over/under voltage* or *over/under frequency*. If either the voltage or the frequency varies from the settings programmed into the inverter, it turns off. This can occur at the end of a utility line or near a manufacturing plant that has lousy power factor. (Power factor was defined in Chapter 4 and is explained in more detail there in the sidebar, "Real Power, Apparent Power, and Power Factor.") In such instances, grid power is poor enough that it doesn't match the window required by the inverter. The inverter remains off until the grid power returns within the acceptable range.

Is My Inverter Safe?

How does a buyer know if his or her inverter will disconnect from the utility during a power outage? It's something utility companies want to know. The answer is on the inverter nameplate. If it lists compliance with UL 1741, you and your utility can rest assured that the inverter will automatically disconnect from the utility grid in case of an over/under voltage or over/under frequency. This feature is most important, of course, when the grid goes down completely (blackout). All grid-tied inverters today are UL 1741 certified. You or your installer will, however, have to inform the utility of this when you file paperwork required to connect to the grid (the interconnection agreement).

Grid-connected inverters also come with a *fault condition reset*—a sensor and a switch that turn the inverter on once the grid is back up or the inverter senses the proper voltage and/or frequency. Here's how it works: When an inverter detects an over/under voltage and/or frequency, it shuts off and remains off for 300 seconds (5 minutes). After 5 minutes, the inverter tests the voltage and frequency of grid power. If it has returned to normal, it will switch back on.

One fact that many who are contemplating a grid-tied solar-electric system often have difficulty accepting is that when the grid goes down or experiences over/ under voltage or over/under frequency, not only does the inverter stop backfeeding surplus onto the grid, the entire wind system shuts down. As a result, the flow of electricity to the main panel and to active branch circuits in the house ceases as well.

As you know by now, the system shuts down to protect utility workers from shock. It also shuts down because the inverter needs the grid connection to determine the frequency and voltage of the AC electricity it produces. Without the connection, the inverter can't operate. It's just wired that way.

To avoid losing power when the grid goes down, you can install a grid-connected system with battery backup. These systems allow homes and businesses to continue to operate when the utility grid has failed. Although inverters in such systems disconnect from the utility during outages, they continue to draw electricity from the battery bank to supply active circuits wired into a critical loads panel. It provides

How Does Power Factor Affect Voltage?

In this chapter you learned that voltage in an AC circuit, when graphed by a special electronic device, forms a perfect sine wave. If you graphed the amperage, you'd find the same thing. For optimum power, the voltage and current waveforms (frequency) should be aligned as closely as possible.

Power factor is a measure of how well the voltage and the current waveforms in electrical current are aligned. The greater their misalignment, the greater the power factor. (For the mathematically inclined, that's because $W = A \times V$.) A 95% power factor, typical of most utility power and grid-tied inverters, means they are pretty close. In some cases, current waveform can lag behind the voltage waveform. For example, large electric motors cause the current waveform to lag behind the voltage waveform. When they are out of alignment, the real power (wattage) delivered to a load decreases.

electricity to the most important (critical) loads. This transition takes place so quickly and smoothly that you won't even know it occurred. In fact, the switch occurs so fast that even computers are not affected.

Another option is the SMA inverter that terminates flow to the main panel and the utility during power outages, but continues to produce electricity so long as the Sun shines.

Grid-connected inverters have LCD displays that provide information on a variety of parameters. The most important are the AC output of the inverter in watts and the number of kilowatt-hours produced by the system during a day. They also provide data on the total number of kilowatt-hours produced by a system since the inverter was installed.

Of lesser importance are the DC voltage and current (amperage) of incoming DC electricity and the voltage of the AC electricity backfeeding onto a grid. (One model, the Fronius inverter seen in Figure 7.7, keeps tab of the dollars saved and pounds of carbon dioxide emissions that you avoided by generating solar electricity.)

Numerous companies produce utility-intertie string inverters, that is, centralized inverters that service one or more strings of PV modules. These include OutBack, Fronius, SMA, Schneider Electric (formerly Xantrex, which was formerly Trace), and SolarEdge. However, when time comes to purchase a wind turbine, you won't have to select an inverter. Wind turbines come with an inverter that's matched to the output of the wind turbine and the controller. The Xzeres 442, for instance, comes with two SMA 6000 inverters. The Xzeres Skystream 3.7 comes with a built-in inverter (it's located inside the turbine along with the generator and other electronics). Some companies, like the US-based Bergey WindPower and Pika Energy, produce their own inverters. They and the controllers that come with the turbine are designed to operate with their wind turbines.

Inverters have long been the weak point in grid-tied wind and solar systems. Fortunately, manufacturers have made changes to enhance performance. Today, string inverters typically come with five- or ten-year warranties.

As a final note, inverters should be installed inside buildings, preferably in dust-free, cool environments. It is true that inverters are rated for outdoor installation

FIGURE 7.7. Utility-intertie Inverter. This sleek, quiet inverter from Fronius is suitable for grid-connected systems without batteries. Fronius's inverters are produced at a new production facility in Sattledt, Austria, which started production in early 2007. The facility receives 75% of its electricity from a large, roof-mounted PV system and 80% of its heat from a biomass heating system. Credit: Fronius.

FIGURE 7.8. Inverter Installed Outside and Inside. Many installers place inverters that service wind turbines outdoors where they are exposed to the elements (*a*). I prefer to install inverters indoors, like the one shown here (*b*). It provides added protection and ensures a longer lifespan. If you do install outdoors, be sure the inverter is shaded from the Sun throughout the year. Credit: Dan Chiras.

and can tolerate cold as well as hot temperatures and moisture; and most installers do place their inverters outside for cost-savings, as shown in Figure 7.8a. However, I always install them in more hospitable environments to ensure a longer productive life. I can't help but think that exposure to extremes in weather as well as rain and snow will shorten an inverter's lifespan. Also, as you shall soon see, inverters operate more efficiently when kept cool.

Off-grid Inverter/Chargers

Off-grid systems and battery-based inverters were once the mainstay of the solar and wind energy business in the United States. Over time, though, as solar and wind electricity entered the mainstream, grid-connected systems have come to dominate the market. In remote areas of Northern Canada, however, off-grid systems are still the mainstay. Fortunately, there are a few companies that produce excellent off-grid inverters (Figure 7.9).

FIGURE 7.9. Battery-based Inverter. This inverter manufactured by Schneider Electric is designed for use on the grid with batteries as well as off grid. For ease of wiring, you can purchase this or similar inverters from other manufacturers pre-wired. That is, they come with all the breakers and internal wiring on the DC and AC side. This makes installation a snap, and saves a lot of brain damage, as wiring these systems can be quite complex. This inverter is about 23 × 16 × 9 inches (58 × 41 × 23 cm) in size. Credit: Schneider Electric.

Like grid-connected inverters, off-grid inverters convert DC electricity into AC and boost voltage to 120 or 240 volts. Many off-grid homes are wired for 120-volt AC since it is impractical to install 240-volt loads like electric dryers, electric stoves, electric water heaters, electric space heat, or ground-source heat pumps. These devices consume way too much electricity for an off-grid system.

Off-grid inverters also perform a number of other functions as well, which we will explore in this section. If you're installing an off-grid system, be sure to read this carefully.

Battery Charging

Battery-based inverters used in off-grid and grid-connected systems with battery backup contain *battery chargers*. (Batteryless grid-tied inverters don't.) As their name implies, battery chargers charge batteries from an external source—either a gen-set in an off-grid system, or the grid in a grid-connected system with battery backup. But isn't the battery charged by the turbine through the charge controller?

In off-grid and grid-tied systems with battery banks, batteries can be charged from the wind turbine or from an external source, either an AC generator or the electric grid. Battery charging from the wind turbine is regulated by the charge controller. It charges batteries with DC electricity. Battery chargers in inverters charge batteries from AC sources. They are a whole different animal. Battery chargers in inverters convert AC to DC at the correct voltage. This is carried out by an electronic circuit containing devices called *rectifiers*. They're simple devices made from diodes that convert AC to DC.

In off-grid systems, battery-charging gen-sets are used to restore battery charge after periods of deep discharge. This prolongs battery life and reduces damage to the lead plates. Battery chargers are also used during equalization, a battery maintenance operation required to increase the life of your batteries. You'll learn more about batteries and both of these processes in Chapter 8.

Battery charging from the electrical grid in a grid-tied system with battery backup occurs after the system has switched to battery power, for example, during a blackout. When the problem's been fixed and the grid is up and running again, the battery bank can be quickly charged using utility power via the battery charger in the inverter.

Abnormal Voltage Protection

High-quality battery-based inverters also contain programmable high- and low-voltage disconnect switches. These features protect various components of a system, such as the batteries, appliances, and electronics. They also protect the inverters themselves.

The low-voltage disconnect (LVD) in an inverter monitors battery voltage at all times. When low battery voltage is detected (indicating that the batteries are deeply discharged) the inverter shuts off (and often sounds an alarm). It no longer produces AC electricity. As a result, the flow of DC electricity from the batteries to the inverter stops. The inverter stays off until the batteries are recharged.

Low-voltage disconnect features are designed to protect batteries from very deep discharging, which can damage batteries. Although lead-acid batteries are designed to withstand deep discharges, discharging batteries beyond the 80% mark causes irreparable damage to the lead plates in a battery and leads to a battery's early demise. Although complete system shut-down can be a nuisance, the disconnect feature is vital to the long-term health of a battery bank.

Although the low-voltage disconnect feature is critical, *Home Power*'s Ian Woofenden notes that "it should not be used to manage one's batteries. It is designed more to protect an inverter from low voltage than to protect batteries from deep discharge. If the voltage of the battery bank reaches the disconnect level," he adds, "you may have already damaged your batteries." It's much better to develop an awareness of the state of charge of your batteries and learn to modify usage or turn on a generator well before the low-voltage disconnect kicks in.

To avoid the hassle of having to manually start a generator when batteries are deeply discharged, some inverters contain a sensor and switch that automatically activate a backup generator. When low battery voltage is detected, the inverter sends a signal to start the generator—provided the generator has a remote start capability. The fossil-fuel generator then recharges the batteries. AC electricity from the gen-set is converted to DC electricity by the inverter's battery charger. It then flows to the batteries, charging them.

When an AC generator is charging batteries, the inverter can send some of the AC electricity to the main service panel (in off-grid systems) or to a critical loads panel (in grid-connected systems) to supply active loads. As you might suspect, more sophisticated auto-start generators cost more than the standard pull-cord type.

High-Voltage Protection

Inverters in battery-based systems also often contain a high-voltage shutoff feature. This sensor/switch terminates the flow of electricity from the gen-set when the battery voltage is extremely high. (Remember: high battery voltages indicate that the batteries are full.) High-voltage protection prevents overcharging, which can severely damage the lead plates in a battery. It also protects the inverter from excessive battery voltage.

Cost Factors

Inverters designed for off-grid systems cost a little more than the less complicated grid-tied inverters. However, the most significant savings in a batteryless grid-connected system comes from the omission of batteries and a battery room. A battery bank with twelve 6-volt batteries could cost upwards of $5,000—and that doesn't include wiring or the construction of a ventilated battery box or battery room.

Another reason off-grid systems cost more is that batteries need replacement every seven to ten years, depending on the type of battery you install and how well you care for them. (If you manage your batteries carefully, you might be able to squeeze 10 to 15 years of service from them.) Deep cycling and allowing the batteries to sit partially discharged for long periods will shorten their lives. Avoiding batteries, therefore, reduces costly battery replacement. It also saves on the time required to maintain batteries, including equalization—a rejuvenating procedure I'll discuss in the next chapter. When comparing costs of an off-grid system to a grid-connected system, don't forget to factor in line extension and utility connection costs for the latter. An off-grid system may be much cheaper than a grid-tied system—if it avoids a $20,000 to $50,000 line extension fee.

Multimode Inverters

Grid-connected systems with battery backup are popular among homeowners who can't afford to be without electricity for a moment or those who experience frequent, long power outages. Grid-tied systems with battery backup require a special type of inverter: a battery- and grid-compatible sine wave inverter. They're commonly referred to as *multifunction* or *multimode inverters*.

Multifunction inverters contain features of grid-connected *and* off-grid inverters. Like a grid-connected inverter, they contain anti-islanding protection, which automatically disconnects the inverter from the grid in case of loss of grid power. They also contain over/under voltage and over/under frequency shutoff. In addition, they contain fault condition reset—to power up an inverter after a problem with the utility grid is rectified. And, like off-grid inverters, multimode inverters contain battery chargers and high- and low-voltage disconnects.

Grid-connected systems with battery backup are the most difficult to understand. Many people think that the batteries are used to supply electricity at night or during periods of excess demand—that is, when household loads exceed the output of a wind turbine. That's not true. In a grid-tied system with battery backup, *the grid* is the source of electricity at night or during periods when demand exceeds the output of the wind turbine. The batteries are there in case the grid goes down. You don't want to use them each night. If the power goes out in the morning, and your batteries have been used heavily at night, they'll be of very little use to you.

When batteries are full, multimode inverters act like any grid-tied inverter. They convert DC electricity from the array into AC that feeds active loads. Surpluses are backfed onto the grid.

If the battery voltage is low, for example, right after a power outage, the inverter charges the batteries and also powers active loads. It can charge the batteries and power AC loads from electricity drawn from the utility lines and, if the wind is blowing, electricity from the turbine. Once the battery bank is full, the system returns to normal operation.

If you are installing an off-grid system, you may want to consider installing a multifunction inverter in case you decide to connect to the grid in the future.

Grid-tied systems with battery backup are not the most efficient systems. That's because a portion of the electricity generated in such systems is used to keep the batteries topped off. (The batteries are trickle-charged to offset self-discharge.) This may only require a few percent of a wind turbine's output, but over time, a few percent adds up. In systems with very old, poorly designed inverters or large backup battery banks, the electricity required to maintain the batteries can be substantial. The amount of energy required to trickle-charge batteries also increases as batteries age. As batteries age, they become less efficient, so more power is consumed to maintain the float charge. (So be careful when purchasing a used inverter. Avoid really old models.)

Buying an Inverter

Inverters come in many shapes, sizes, and prices. The smallest inverters, referred to as *pocket inverters*, range from 50 to 200 watts. They are ideal for small loads such as VCRs, computers, radios, televisions, and the like. Larger units range from 1,000 to 5,500 watts. Most home and small businesses require inverters in the 2,500 to 5,500-watt range. Which inverter should you select?

Type of Inverter

The first consideration when shopping for an inverter is the type of system you are installing. As noted earlier, if you are installing an off-grid system, you'll need a battery-based inverter. You will also need to purchase a wind turbine designed for battery charging—one that produces DC electricity at the same voltage as your battery bank. The wind turbine will also come with a controller.

If you are installing a batteryless grid-connected system, you'll need a turbine designed for this purpose and its matching inverter. As noted earlier, the inverter must be matched to the output of the wind generator. Inverters are usually sold with the turbine to ensure compatibility. If not, the manufacturer will sell you the wind machine and controller and will recommend a compatible inverter. Follow their advice.

If you are installing a grid-connected system with battery backup, purchase an inverter that is both grid- and battery-compatible. Battery-based grid-tied inverters may also be a good choice for off-grid systems, in case you decide to connect to the electrical grid at a later date.

When purchasing an inverter through an installer, your choices will very likely be limited. Most installers install inverters in which they have a high degree of confidence. An installer may also be a dealer for one or two wind turbines, each of which comes with its own inverter. If this is the case, the installer will sell you what you need. I should point out that there are not that many battery-based inverters available in North America.

Even though a local supplier/installer may recommend an inverter, I urge you to read the following material. It will expand your understanding of inverters, and will help you better manage and troubleshoot your system. If you are designing and installing your own system, which I recommend only after serious study and a couple of hands-on workshops, you'll surely want to know as much as you can about inverters. Here's what you need to know and consider as you shop for a battery-based inverter.

System Voltage

When shopping for a battery-based inverter, you'll need to select one with an input voltage that corresponds with the battery voltage of your system. System voltage refers to the voltage of the electricity produced by a renewable energy technology— a wind turbine and its controller, a solar-electric array, or micro hydro generator.

In a wind system, the turbine with its controller is designed to produce either 12-, 24-, or 48-volt electricity. The batteries are wired similarly. Inverters, in turn, increase the voltage to the 120 or 240 volts AC used in homes and businesses. Because components of an off-grid wind energy system must operate at the same voltage, the inverter and its controller must match the batteries. A 24-volt inverter won't work in a 48-volt system. If you are installing a 48-volt Bergey XL-R, you'll need a 48-volt battery-based inverter, and you must wire your battery bank for 48-volts. If installing a 24-volt Kestrel 1000, you'll need a 24-volt battery-based inverter and battery bank. It is always a good idea to talk with the wind turbine manufacturer's technical staff to obtain their input on the best inverter.

Wind Turbine Options

Wind machines designed for battery charging often come in one of three standard voltages: 12, 24, and 48 volts. The 1-kW Bergey XL, for example, is only available in 12-, 24- and 48-volt models. Others come in numerous voltages designed to meet individual needs. The Bergey 7.5 Excel (Formerly Excel R), for example, is available in 24-, 48-, 120-, and 240-volt DC battery-charging configurations. The Kestrel e400, manufactured by Eveready, is also available in numerous voltages (48, 110, and 250 volts DC).

Waveform: Modified Square Wave vs. Sine Wave

The next inverter selection criterion you must consider is the output waveform. Off-grid inverters are available in two types: modified square wave (often called *modified sine wave*) and sine wave. Grid-connected inverters are all sine wave so they match utility power. Multimode inverters for grid-tied systems with battery backup must be sine wave. (You can't backfeed onto an electrical grid unless you are sending sine wave electricity onto it.)

As you may recall, modified square wave electricity is a crude approximation of grid power, but it works well in many appliances such as refrigerators and washing machines and in power tools. It also works pretty well in most electrical devices, including TVs, lights, stereos, computers, and inkjet printers. Although all these devices can operate on this lower-quality waveform, they all run less efficiently, producing more heat and less work—light or water pumped, etc.—for a given input.

Problems arise, however, when modified square wave is fed into sensitive electronic circuitry such as microprocessor-controlled front-loading washing machines; appliances with digital clocks; chargers for various cordless tool; copiers; and laser printers. These devices require sine wave electricity. Without it, you're sunk. I, for example, found that my energy-efficient front-loading Frigidaire Gallery washing machine would not run on the modified square wave electricity produced by my very first inverter. The microprocessor that controls this washing machine—and other similar models (except the German-made Staber)—simply can't operate on this inferior form of electricity. After I replaced the inverter with a sine wave inverter, I had no troubles whatsoever.

Certain laser printers may also perform poorly with modified square wave electricity. The same goes for some battery tool chargers, ceiling fans, and dimmer switches.

Making matters worse, some electronic equipment, such as TVs and stereos, give off an annoying high-pitched buzz or hum when operating on modified square wave electricity. Modified square wave electricity may also produce annoying lines on TV sets, and can even damage sensitive electronic equipment.

When operated on modified square wave electricity, microwave ovens cook more slowly. Equipment and appliances also run warmer and might last fewer years. Computers and other digital devices operate with more errors and crashes. The only time I burned out a computer was when I was powering my off-grid home with a modified square wave inverter, and I've been working on computers since 1980. In addition to these problems, digital clocks don't maintain their settings as well when operating on modified square wave electricity. Moreover, motors don't always operate at their intended speeds. So why do manufacturers produce modified square wave inverters?

The biggest reason is cost. Modified square wave inverters are much cheaper than sine wave inverters. You will most likely pay 30 to 50 percent less for one.

Another reason for the continued production of modified square wave is that they are hardy beasts. They work hard for many years with very little, if any, maintenance. (Their durability may be related to their simplicity: they are electronically less complex than sine wave inverters.)

Modified square wave inverters come in two varieties: high-frequency switching and low-frequency switching units. High-frequency switching units are the cheaper of the two. A typical 2,000-watt high-frequency switching inverter, for example, costs 20 to 50 percent less than low-frequency models. They are also much lighter than low-frequency switching models, and are, therefore, easier to install. A high-frequency inverter may weigh 13 pounds compared to the 50 pounds of a low-frequency inverter. However, it is the latter that is used in wind systems.

Although low-frequency switching modified square wave inverters cost more and weigh more than their high-frequency switching cousins, they are well worth the investment. One reason for this is that they typically have a much higher surge capacity. That means they can deliver the greater surges of power that are needed to start certain electrical devices such as well pumps, power tools, dishwashers, washing machines, and refrigerators. More on surge capacity shortly.

Although there are some reliable modified square wave inverters on the market, I recommend that you purchase a sine wave battery-based inverter for off-grid systems. Their output is well suited for use in modern homes with their array of sensitive electronic equipment. SMA, Schneider Electric, and OutBack all produce excellent sine wave inverters.

Continuous Output

Continuous output is a measure of the power an inverter can produce on a continuous basis—provided there's enough energy available in the system. The power output of an inverter is measured in watts, although some inverter spec sheets also list continuous output in amps (to convert watts to amps use the formula watts = amps × volts). Schneider Electric's (formerly Xantrex) sine wave inverter, model SW2524, for instance, produces 2,500 watts of continuous power. This inverter can power a microwave using 1,000 watts, an electric hair dryer using 1,200 watts, and several smaller loads simultaneously without a hitch. (By the way, the "25" in the model number indicates the unit's continuous power output; it stands for 2,500 watts. The "24" indicates that this model is designed for a 24-volt wind system.) The spec sheet on this inverter lists the continuous output as 21 amps.

OutBack's sine wave inverter VFX3524 produces 3,500 watts of continuous power and is designed for use in 24-volt systems. Off-grid homes can easily get by on a 3,000- to 4,000-watt inverter. Homes rarely require the full output.

To determine how much continuous output you'll need, add up the wattages of the common appliances you think will be operating at once. Be reasonable, though. Typically, only two or three large loads operate simultaneously. However, remember that, in some instances, multiple loads can operate simultaneously. A washer and well pump may be operating, for example, when a sump pump is running. If you are planning to operate a shop next to your home on the same inverter, you will need an inverter with a higher continuous output.

Surge Capacity

Electrical devices with motors, such as vacuum cleaners, refrigerators, washing machines, and power tools, require a surge of power to start up. To observe this, simply watch the amp meter in a renewable energy system when someone turns on a device such as a vacuum cleaner. Or watch the amperage on a Watts Up? or Kill A Watt meter, described in Chapter 4. If you pay close attention, you'll see a momentary spike in amperage. That's the power surge required to get the motor running. The spike typically lasts only a fraction of a second. Even so, if an inverter doesn't provide a sufficient amount of surge power, the power tool or appliance won't start! Not starting is not just inconvenient. Stalled motors draw excessive current and overheat very quickly. Unless they are protected with a thermal cutout, they may burn out.

When shopping for an inverter, be sure to check out surge capacities. All quality low-frequency inverters are designed to permit a large surge of power over a short period, usually about five seconds. Surge capacity or surge power is listed on spec sheets in watts and/or amps.

Efficiency

Another factor to consider when shopping for inverters is their efficiency. Efficiency is calculated by dividing the energy coming out of an inverter by the energy going in.

While inverter efficiencies range from 80% to 95%, most models boast efficiencies in the mid to low 90s. A 94% efficient inverter, for instance, loses 6% of the energy it converts from DC to 120-volt AC.

It should be noted that the efficiency of inverters varies with load. Generally, an inverter achieves its highest efficiency once output reaches 20% to 30% of its rated capacity, according to Richard Perez, renewable energy expert and publisher of *Home Power* magazine. A 4,000-watt inverter, for instance, will be most efficient at outputs above 800 to 1,200 watts. At lower outputs, efficiency is dramatically reduced.

Cooling an Inverter

Because they're not 100% efficient, inverters produce waste heat internally. A 4,000-watt inverter running at the rated output at 90% efficiency produces 400 watts of heat internally. (That's equivalent to the heat produced by four 100-watt incandescent lightbulbs.) The inverter must get rid of this heat to avoid damage.

Inverters rely on cooling fins to passively rid themselves of excess heat to reduce internal temperature. Fins increase the surface area from which heat can escape. Some inverters also come with fans to provide active cooling. Fans blow air over internal components, stripping off heat. If the inverter is in a hot environment, however, it may be difficult for it to dissipate heat quickly enough to maintain a safe temperature.

Unbeknownst to many, inverters are also often equipped with circuitry that protects them from excessive temperature. That is, they're programmed to power down (produce less electricity) when their internal temperature rises above a certain point. As the internal temperature of the power oscillator (H-bridge) in a transformer-based inverter increases, the output current decreases. Translated, that means if you need a lot of power, and the internal temperature of your inverter is high, you won't get it. A Schneider Electric SW series inverter, for example, produces 100% of its continuous power at 77°F (25°C), but drops to 60% at 117.5°F (47.5°C). Although this sounds like an unusually high temperature, remember that an inverter produces its own internal heat. It can raise internal temperature well above ambient temperature.

The implications of this are many. First, as I've pointed out before, inverters should be installed in relatively cool locations. If the inverter is mounted outside, make sure it is shaded. Second, inverters should be installed so that air can move freely around them. Don't box an inverter in to block noise (though not all inverters are noisy).

Battery Charger

Battery chargers are standard in battery-based inverters. As you may recall, a battery charger will allow you to charge your DC battery bank using AC from the utility (for grid-connected systems with battery backup) or an AC generator (for either off-grid systems or grid-connected systems with battery backup).

Noise and Other Considerations

Inverters can be installed inside a home or outside. Most inverters are rated NEMA-3R. NEMA stands for National Electrical Manufacturers Association; 3R is the level of weather protection. It means the inverters are contained in weather-tight enclosures that can withstand rain, snow, and even ice forming on their surface. Even though inverters are rated for outdoor installation, I believe they should be protected from very low and very high temperatures. In fact, many grid-tied inverters are rated for installations with ambient air temperatures above −10°F.

Battery-based inverters are typically installed inside so they can be located close to the batteries, but also for weather protection. (As noted, the batteries also need to "live" in a warm space.) Wires from battery banks carry either 12, 24, or 48-volt DC. To reduce line loss, it is best to minimize line runs. This is especially true because Code requires the use of very large wires, which are expensive.

When installed inside, grid-tied inverters are almost always installed near the main panel. That's where the utility service enters a house. (Most inverter manufacturers recommend their equipment be housed at room temperature.)

If you are planning on installing an inverter inside your business or home, be sure to check out the noise it produces. Inquire about this upfront, or, better yet, ask to listen to the model you are considering to be sure it's quiet. Don't take a manufacturer's word for it. My first inverter was described by the manufacturer as "quiet," but quiet compared to a jet engine. It emitted an annoyingly loud buzz heard throughout my house. The first six months after I moved into my off-grid, rammed-earth tire home, the inverter's buzz drove me nuts, but I grew used to it.

Some folks are also concerned about the potential health effects of extremely low-frequency electromagnetic waves emitted by inverters as well as other electronic equipment and electrical wires. If you are concerned about this, install your inverter in a place away from people. Avoid locations in which people will be spending a lot of time—for example, don't install the inverter on the other side of a wall from your bedroom or office.

While creating a checklist of features to consider when purchasing an inverter, be sure to add ease of programming. My very first Trace inverter (DR2424, modified square wave) was a dream when it came to programming: all of the controls were manually operated dials. To change a setting, all I had to do was turn the dial. Its replacement, a Trace PS2524, which was a sine wave inverter, worked wonderfully but was challenging to program. Digital programming can be extremely complicated, and the instructions can be difficult to follow.

My advice is to find out in advance how easy it is to change settings—and don't rely on the opinions of salespeople or renewable energy geeks that can recite pi to the 20th decimal place. Ask friends or dealers/installers for their opinions, but also ask them to show you how it's done. You may even want to spend some time with the manual (often available online) to see if it makes sense *before* you buy an inverter.

Another feature to look for in a battery-based inverter is power consumption under search mode. What's that?

The search mode is an operation that allows a battery-based inverter to go to sleep, that is, shut down almost entirely in the absence of active loads. The search mode saves energy (inverters may consume 10–30 watts or more from the battery when in the "on" mode).

Although the inverter is sleeping when it is in search mode, it's sleeping with one eye slightly open. That is, it's on the alert should someone switch on a light or an appliance. It's able to do this by sending out tiny pulses of electricity through the electrical wires approximately every second, searching for an active load. When an appliance or light is turned on, the inverter senses the load and quickly snaps into action, powering up and feeding AC electricity.

The search mode is handy in houses in which phantom loads have been eliminated. (As you may recall, a phantom load is a device that continues to draw a small electrical current when off.) According to the US Department of Energy, phantom loads, on average, account for about five percent of a home's annual electrical consumption. (In some homes, however, they can be as high as ten percent.) Eliminating phantom loads saves a small amount of electrical energy in an ordinary home. In a home powered by renewable energy, however, it saves even more because supplying phantom loads 24 hours a day requires an active inverter. An inverter may consume about 30 watts when operating at low capacity. Servicing a 12-watt phantom load, therefore, requires an additional 30-watt investment in energy in the inverter—24 hours a day.

Energy savings created by eliminating phantom loads and continuous inverter operation does have its downsides. For example, automatic garage door openers may have to be turned off for the inverter to go to into the search mode. In my off-grid home in Colorado, I turned off my garage circuit at the breaker box or main service panel when I'd leave home. It was a pain in the neck, however, because when

I came home, I'd have to switch it back on to open the garage door. A smarter way to handle this might be to put the garage door opener on a circuit controlled by a switch located outside the garage.

Another problem occurs with electronic devices that require tiny amounts of electricity to operate, like cell phone chargers. When left plugged in, cell phone chargers may draw enough power to cause an inverter to turn on. Once the inverter starts, however, the device doesn't draw enough power to keep the inverter going. As a result, the inverter switches on and off, *ad infinitum*. I found the same with my portable stereo.

Because of these problems, many people simply turn the search mode off so that the inverter keeps running 24 hours a day. Another option is to set the search mode sensitivity up, so it turns on at a higher wattage. However, most modern homes have at least one always-on load, for example, hard-wired smoke detectors or garage door openers that require the continuous operation of the inverter.

Computer Interfacing and Monitoring

When I wrote the first edition of this book, just a few years ago, only the more sophisticated—and most expensive—inverters came with remote monitoring. That is, they had electronics that kept track of energy production and could send that information to a computer or to a website that could be accessed by computer.

Today, virtually all grid-tied inverters come with a data logging capability. This data can be uploaded to a computer or to a website. Most commonly, data is stored and displayed online at the manufacturer's website. However, you can subscribe to third-party monitoring services. (I've listed them in the Resource Guide at the end of the book.) As a result, a homeowner can check up on his or her system anywhere in the world that has internet access. For example, when you are vacationing in Mexico, you can go online to see how your system is performing. (Although you really should be swimming in the ocean or enjoying a margarita.)

Remote monitoring is a lot of fun but also quite useful because it allows homeowners and installers an opportunity to monitor renewable energy systems for

FIGURE 7.10. Power Panel with Four Inverters Wired in Series. In olden days, installers often stacked inverters to increase the voltage and amperage in battery-based renewable energy systems. Two inverters, for instance, were often stacked to produce 240-volt AC electricity. Those days have largely ended because inverter manufacturers have created battery-based inverters that produce 240-volt electricity without stacking. A single inverter, for instance, can be programmed to produce either 120-volt AC for off-grid applications or 240-volt AC for grid-tied homes (with battery backup). Credit: Outback Power.

potential problems. I go online every month or so to check the systems I've installed for my solar customers. Online, you can see the output of each system at any given moment and over time. Similar monitoring is available for wind systems.

Other monitoring systems are available, but they are all basically the same. They consist of data loggers and communications gateways. Even though I install remote monitoring on most of my systems, I always install a separate customer-owned utility meter as a backup on solar and wind energy system. (Utility companies will often give you old utility meters that you can install yourself for your own use.) This meter is placed between the inverter and the main panel. It keeps track of the wind turbine's cumulative output. That way, if anything goes wrong with the computer-based monitoring system, you will know what your system has produced. It's come in handy many times in my career.

Stackability

In the first edition of this book, I recommended that when buying a battery-based inverter a homeowner select one that can be stacked—that is, connected to a second or third inverter of the same kind in parallel or series. That's because the original battery-based inverters were small. They only produced 120-volt AC electricity. If you wanted 240-volt AC, you had to wire their AC output in series. Figure 7.10 shows four 120-volt OutBack inverters wired in a combination of series and parallel.

Although 120-volt inverters are still on the market, manufacturers now produce impressive inverters that produce either 120 or 240. There's no need to stack them.

Stackability and the 120/240-volt inverters allow one to produce 240-volt electricity to operate appliances such as electric clothes dryers, electric stoves, central air conditioning, or electric resistance heat. As a general rule, I recommend that you avoid such appliances if you're considering an off-grid system. That's not because a wind system can't be designed to power these loads, but rather because they tend to use lots of electricity, and you'll need to install a much larger—and much more costly—wind turbine to power them.

Well-designed, energy-efficient homes can usually avoid using 240 volts AC. A frequent exception is a deep well pump, which may require 240 volts, but in most cases, a high-efficiency 120-volt AC pump, or even a DC pump, can do the job. Several companies manufacture 120/240 volt inverters, including Magnum Energy, OutBack, and Schneider Electric.

Another option—for off-grid homes that require only one 240-volt outlet—is to install a step-up transformer to supply one or more 240-volt AC outlets. These units step up the 120-volt AC electricity from an inverter to 240-volt AC electricity.

Conclusion

If you are hiring an honest, experienced professional to install a wind energy system, you won't have to worry about buying the right inverter. When installing your own system, be sure that your source takes the time to determine which inverter is right for you. Provide as much information about your needs as possible, and ask lots of questions.

A good inverter is key to the success of a renewable energy system, so shop carefully. Size it appropriately. Be sure to consider future electrical needs. But don't forget that you can trim electrical consumption by installing energy-efficient electronic devices and appliances. Efficiency is always cheaper than adding more capacity. When shopping, select the features you want and buy the best inverter you can afford. Although modified square wave inverters work for most applications, it is best to purchase a sine wave inverter.

CHAPTER 8

Batteries, Charge Controllers, and Backup Generators

Batteries are the unsung heroes of off-grid renewable energy systems. Even though wind turbines receive most of the attention, work hard, and endure extreme weather, the batteries in off-grid systems operate day in and day out, 365 days a year for up to ten years, give or take a little, provided you treat them well. Although batteries in off-grid systems do indeed work hard, these brutes of the renewable energy field require considerable pampering. You can't simply plop them in a battery room, wire them into your system, and then go about your business expecting your batteries to perform at 100% of their capacity for the lifetime of your system. You'll need to install them properly, watch over them carefully, and tend to their needs. If you don't, they'll die young, many years before their time.

Because batteries are so important to off-grid renewable energy systems and require so much attention, it is essential that readers who are thinking about installing an off-grid system or a grid-connected system with battery backup understand a fair amount about batteries. Doing so will help you get the most from your batteries and could save you a fortune over the long haul. This chapter will provide a great deal of essential information about batteries and also help readers develop a solid understanding of two additional components typically required in battery-based systems: charge controllers and generators. They are vital to the health and vitality of a battery bank.

If you are only interested in a batteryless grid-connected system, you don't need to read this chapter. Feel free to skip to Chapter 9.

Understanding Lead-Acid Batteries

Batteries are a mystery to many people. How they work, why they fail, and what makes one type different from another are topics that can boggle the mind. Fortunately, the batteries used in most off-grid renewable energy systems are pretty much the same: they're deep-cycle, flooded lead-acid batteries.

Flooded lead-acid batteries used in most renewable energy systems are the ultimate in rechargeable batteries. They can be charged and discharged (cycled) a thousand or more times before they wear out, provided you take good care of them.

Lead-acid batteries used in most renewable energy systems contain three cells—that is, three distinct 2-volt cells or compartments. Inside the battery case, the individual cells are electrically connected (wired in series). As a result, they collectively produce 6-volt electricity. (Wiring in series, that is, positive to negative, increases voltage.)

Inside each cell in a flooded lead-acid battery is a series of thick, parallel lead plates, as shown in Figure 8.1. Each cell is filled with battery fluid, hence the term "flooded." Battery fluid consists of 70% distilled water and 30% sulfuric acid. A partition wall separates each cell, so that fluid cannot flow from one cell to the next. The cells are enclosed in a heavy-duty plastic case.

As illustrated in Figure 8.1, two types of plates are found inside a battery: positive and negative. They are wired in series inside the battery and connected to the

The Life of a Battery

Off-grid systems require batteries. Without them, there would be no electricity during the periods when a wind turbine or PV array is idle. Batteries ensure a steady supply of electricity to off-gridders. When the wind is not blowing or when demand exceeds output of the wind turbine array, batteries are there to provide the electricity needed to keep homes running smoothly. When surplus electricity is available, the batteries store it for later use.

While batteries in an off-grid system work day and night, batteries in a grid-connected system with battery backup live a life of leisure. Their main function is to store electrical energy in case of an emergency. In between power outages, however, the batteries lounge in the comfort of the battery room, filled to capacity with electricity. Their sole purpose is to remain fully charged in case they are called into duty.

As a general rule, the largest (and most costly) battery banks are required for off-grid systems. Wind/PV hybrid systems require slightly smaller battery banks, for reasons explained in Chapter 3. Grid-connected wind systems with battery backup generally require even smaller battery banks—typically just enough to get a family or business by for a few hours or a few days should grid power fail.

battery posts or terminals. Battery posts allow electricity to flow into and out of a battery.

As shown in Figure 8.1, the positive plates of lead-acid batteries are made from lead dioxide (PbO_2). The negative plates are made from pure lead. Sulfuric acid fills the spaces between the plates and is referred to as the *electrolyte*. It plays an extremely important role in batteries, and will be discussed shortly.

How Lead-Acid Batteries Work

Although battery chemistry can be a bit daunting to those who have never studied the subject or had it explained well, it's important that all users understand a little about the chemical reactions.

Like all other types of batteries, lead-acid batteries convert electrical energy into chemical energy when charging. When discharging (that is, giving *off* electricity), chemical energy is converted back into electricity. How does this occur?

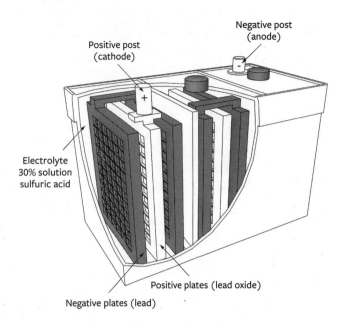

FIGURE 8.1. Cross Section of Lead-Acid Battery. Credit: Anil Rao.

As shown in Figure 8.2, when electricity is drawn from a lead-acid battery, sulfuric acid reacts chemically with the lead of the negative plates. As noted in the top equation, this chemical reaction yields electrons, tiny negatively charged subatomic particles. The electrons flow out of the battery via the negative terminal through the battery cable.

During this chemical reaction, you will note that lead reacts with sulfate ions to form lead sulfate. As a result, tiny lead sulfate crystals form on the surface of the negative plates when the battery is discharging.

The second equation in Figure 8.2 shows the chemical reaction that takes place on the surface of the positive plates when a battery is discharging; sulfuric acid reacts with the lead dioxide of the positive plates, forming tiny lead sulfate crystals on them as well. Keep that in mind, because it's the formation of lead sulfate that reduces a battery's lifespan. Excessive buildup can destroy a battery. Your job as an off-grid solar system operator is to see to it that lead sulfate is quickly removed

FIGURE 8.2. Chemical Reactions in a Lead-Acid Battery. The chemical reactions occurring at the positive and negative plates are shown here. Note that these reactions are reversible. Lead sulfate forms on both the positive and negative plates during discharge. Credit: Anil Rao.

$$Pb(s) + HSO_4^-(aq) \rightarrow PbSO_4(s) + H^+(aq) + 2e^- \qquad \text{negative plates}$$
$$PbO_2(s) + HSO_4^-(aq) + 3H^+(aq) + 2e^- \rightarrow PbSO_4(s) + 2H_2O(1) \qquad \text{positive plates}$$

A Word of Warning

Sulfuric acid is a *very* strong acid. In fact, it is one of the strongest acids known to science. In flooded lead-acid batteries, sulfuric acid is diluted to 30%. Although diluted, it is still to be treated with great respect—it can burn your skin and eyes and eat through clothing like a ravenous moth. Be sure to wear eye protection and rubber gloves when filling batteries. It's a good idea to remove jewelry and wear a long-sleeved shirt you don't care about. It's bound to get a little acid on it. Acid will produce tiny holes in the shirt that you won't notice until you wash it.

from the plates; you do it by reversing the chemical reaction—that is, charging the batteries as soon as possible after discharging.

Discharging a battery not only creates lead sulfate crystals, it converts some of the sulfuric acid into water. However, as you shall soon see, the chemical reactions just described are reversible. Therefore, when a battery is charged (when electricity is delivered to the battery), the chemical reactions at the positive and negative plates run in reverse. As a result, lead sulfate crystals on the surface of both the positive and negative plates are broken down and sulfuric acid is regenerated. The plates are restored, although a small number of lead sulfate crystals may flake off, slowly but surely whittling away at the plates.

Although the chemistry of lead-acid batteries is complicated, what is most important to remember is that these chemical reactions allow us to store electricity (electrons) inside batteries and reclaim them when we need electricity.

Will Any Type of Lead-Acid Battery Work?

Lead-acid batteries are used in a wide range of applications. For example, there's a 12-volt flooded lead-acid battery under the hood of your automobile. Trucks, tractors, and buses have similar batteries, only larger. In vehicles, batteries provide the electricity required to start the engine and operate lights, the ignition system, and accessories like a clock or a radio when the engine's not running. As a result, these batteries are known as *starting, lighting, and ignition*, or *SLI*, batteries.

Lead-acid batteries are also used in a wide assortment of electric vehicles, including forklifts, golf carts, and electric lawn mowers (Figure 8.3). They are even used in emergency standby power systems, providing backup power in case the electrical grid goes down. Lead-acid batteries are also used in RVs, sailboats, yachts, and powerboats, although sealed batteries are typically used in these applications so they won't spill acid when the boat or RV rocks. (More on sealed batteries shortly.)

Lead-acid batteries come in many varieties, each one designed for a specific application. Car batteries, for example, are designed for use in cars, light trucks, and vans. Their thin, porous plates offer lots of surface area to the electrolyte. This allows the battery to discharge large numbers of electrons when you start your car.

This high-amp current enables the battery to turn the starter motor, an event that requires a huge amount of current.

Unfortunately, the thin lead plates of an SLI battery can easily be damaged by deep discharges. That's because a lot of the lead sulfate that forms on the plates flakes off, thinning the plates. Leave your car's headlights on a few times and you'll destroy the battery. Marine batteries are similar but are optimized for use in boats—for starting engines and providing small amounts of electricity to power electrical equipment like radios and GPS units. That said, there are deep-cycle marine batteries on the market, too. I'll discuss them shortly.

While most lead-acid batteries—even car batteries—can be installed in a renewable energy system, for optimum performance and years of trouble-free service, I highly recommend using batteries designed specifically for renewable energy systems: deep-cycle, flooded lead-acid batteries (Figure 8.4). They can be safely deep discharged over and over again with relative impunity, for reasons explained shortly.

Even though batteries designed for deep discharge are considerably more robust than car and truck batteries, they are not invincible. If a battery is discharged too deeply—more than 80% of its capacity—it will be permanently damaged. Fortunately, controls in a well-designed renewable energy system prevent this from occurring.

FIGURE 8.3. My Electric Truck. A couple friends and I converted this Chevy S-10 to operate on electricity. We removed the gas engine and all internal combustion-related components, like the exhaust system and gas tank, and replaced them with an electric motor and 24 deep-cycle golf cart batteries. This truck can travel a little under 60 miles on a single charge and easily cruises at 55 to 60 miles per hour (88 to 97 kilometers per hour). Credit: Dan Chiras.

FIGURE 8.4. Lead-Acid Batteries for RE System. Lead-acid batteries are commonly used in renewable energy systems. Be sure to purchase deep-cycle batteries; keep them in a warm, safe place; recharge quickly after deep discharges; and equalize them periodically. Credit: Surrette Battery Company LTD.

For optimum performance, deep-cycle batteries also need to be recharged fairly soon after undergoing a discharge—even if they are only discharged 20% to 30% of their capacity. The reason for this is that if the batteries are left in a state of discharge for an extended period, lead sulfate crystals begin to coalesce. Small crystals become large. These crystals "insulate" the electrolyte from the lead plates. This reduces the ability of the battery to store electricity. Because batteries take progressively less charge, they have less to give back. Making matters worse, large lead sulfate crystals are also not completely converted back to lead or lead dioxide when you charge a battery that's been left in a state of deep discharge for an extended period. Over time, then, entire cells in a battery may die, effectively ending a battery's useful life.

To avoid lead sulfate crystal growth, batteries should be promptly recharged and also periodically *equalized*. This process, discussed shortly, helps rejuvenate batteries by driving lead sulfate off the plates. It also equalizes the voltage of the cells in a battery bank.

So, do not forget: *For optimum long-term performance, batteries need to be recharged as quickly as possible after deep discharging to ensure a long, productive life.* Deep-cycle, flooded lead-acid batteries like the industry standard L16s, should last for seven to ten years, maybe even longer if they are promptly recharged after deep discharging, routinely equalized, and housed in a warm location. If you fail to perform this routine maintenance and make other mistakes like not filling the batteries with distilled water when needed, you can count on having to write a very large check for a replacement battery bank much sooner than necessary.

Can I Use a Forklift, Golf Cart, or Marine Battery?

Forklifts require huge, high-capacity, deep-discharge batteries. These leviathans can be used in renewable energy systems because they are designed for a long life and operate under fairly demanding conditions. In fact, they can withstand 1,000 to 2,000 deep discharges—more than standard deep-cycle batteries typically used in battery-based renewable energy. This makes them ideally suited for off-grid wind and solar energy systems.

Although forklift batteries function very well in such instances, they are rather heavy and expensive. If you can acquire them at a decent price, you might want to use them, especially if the size of your system warrants a large battery. You will, however, probably need to rent a forklift to install them in your battery room.

Golf cart batteries can also work. Like forklift batteries, golf cart batteries are designed for deep discharge. Moreover, they typically cost a lot less than larger deep-cycle batteries. While the lower cost may be appealing, remember that there's a reason for this: golf cart batteries are smaller than standard batteries (L16s, for instance) used in renewable energy systems and, therefore, don't store as much electricity. They also don't last as long. In fact, they may last only five to seven years—if well cared for. Shorter lifespan means more frequent replacement. More frequent replacement means higher long-term costs and more work on your part.

Although golf cart batteries don't last as long as the heavier-duty, deep-cycle batteries, some renewable energy system installers swear by them. Others recommend them for first-time users, who are bound to abuse their first set of batteries. They're a training set. If you screw up and kill your battery bank, you won't be out as much money.

Another option many people consider, because they are widely available, are deep-cycle marine batteries. Unfortunately, even though they are advertised as deep-cycle or "marine deep-cycle" batteries, they are a compromise between a car starting battery and a deep-cycle battery. Their thinner plates just aren't up to the task of a renewable energy system. They will not last as long in deep-cycle service as true deep-cycle batteries.

What About Used Batteries?

Although used batteries can sometimes be purchased for pennies on the dollar, and may seem like a bargain, for the most part, they're not worth it. As a buyer, you have no idea how well—or more likely, how poorly—they've been treated. You can test the voltage to determine whether they are functional, but there are questions you may not be able to find answers to. For example, how old are they? Have they been deeply discharged many times? Have they been left in a state of deep discharge for long periods? Have they been filled with tap water rather than distilled water? Have they been stored in a cold environment?

Many used batteries are discarded because they've failed or are experiencing a serious decline in function. As noted, with age comes decreased efficiency that translates into decreased storage capacity. If a used battery is 80% efficient, you have to put 120 amp-hours into the battery for every 100 amp-hours you take out. If the battery is only 50% efficient, you have to put 200 amp-hours into it for every 100 amp-hours you draw out.

There are some exceptions to my advice not to buy used batteries. For example, you may be able to find high-quality, gently used batteries from telephone

Don't Skimp on Batteries

When shopping for batteries for a renewable energy system, I recommend that you buy high-quality deep-cycle batteries. L16s are an industry standard. Although you might be able to save some money by purchasing a cheaper alternative, frequent replacement is a pain in the neck and can be costly. Batteries are heavy, and it takes quite a lot of time to disconnect old batteries and rewire new ones. Bottom line: the longer a battery will last—because it's the right battery for the job, and it's well made and well cared for—the better.

FIGURE 8.5. Sealed Lead-Acid Battery. Sealed batteries like these 12-volt batteries never need to be watered. Nonetheless, they still require proper housing and temperature conditions and careful control of state of charge to ensure a long life. Notice the battery posts and nonremovable pressure-release caps. Credit: Dan Chiras.

companies. They routinely replace batteries in their repeater stations (stations for landlines that amplify signals and transmit them on to the next station). These batteries may be a good bargain.

Whatever your source of used batteries is, shop carefully. Find out how the batteries were used. And remember, although there are exceptions, most people who've purchased used flooded lead-acid batteries have been disappointed.

Sealed Batteries

While most installers of off-grid wind systems recommend flooded lead-acid batteries, they often prefer to install "sealed" lead-acid batteries in grid-connected systems with battery backup. Sealed batteries are also known as *captive electrolyte batteries*. They are filled with electrolyte at the factory, charged, and then permanently sealed. Because they are sealed, they are easy to handle and ship without fear of spillage. They won't even leak if the battery casing is cracked open, and they can be installed in any orientation—even on their sides. If you are pressed for space in your battery room, these batteries might work well.

Unlike flooded lead-acid batteries, which require periodic filling with distilled water, sealed batteries never need watering. In fact, you *can't* add distilled water. There are no fill-caps (Figure 8.5).

Two types of "sealed" batteries are available: *absorbed glass mat* (AGM) batteries and *gel cell* batteries. In absorbed glass mat batteries, thin absorbent fiberglass mats are placed between the lead plates. The mats consist of a network of tiny pores that immobilize the battery acid—so it won't spill out if a battery is carried, jostled (as in a sailboat), or laid on its side. The fiberglass mats also create tiny pockets that capture hydrogen and oxygen gases given off by the lead plates during charging. Hydrogen comes from two sources. First is the chemical reactions that occur during charging. Second is hydrolysis, the breakdown of water. As a battery charges, some of the electricity splits water molecules into hydrogen and oxygen. In sealed batteries, hydrogen and oxygen cannot escape. There are no caps to allow it to vent. These gases accumulate in the tiny pockets in the glass matt where they react with each other, forming water. (In flooded lead-acid batteries, these gases escape through small openings in the battery caps. It's because of hydrolysis that you need to add distilled water every month to a flooded lead-acid battery.)

> ## Sealed Batteries—A Misnomer?
>
> The truth be known, sealed batteries are not totally sealed. Each battery contains a pressure-release valve that allows gases and fluid to escape if a battery is accidentally overcharged. The valve keeps the battery from exploding. Once the valve has blown, however, the battery will very likely need to be retired.

In gel batteries, the sulfuric acid electrolyte is converted to a substance much like hardened Jell-O by the addition of a small amount of silica gel. The gel-like substance fills the spaces between the lead plates and eliminates liquid electrolyte, much like the fiberglass mats in an AGM battery.

Sealed batteries are often referred to as *maintenance-free* batteries because fluid levels never need to be checked and because the batteries never need to be filled with distilled water. They also never need to be (and should not be!) equalized, a process discussed later in this chapter. Eliminating routine maintenance saves a lot of time and energy. It makes sealed batteries a good choice for off-grid systems in remote locations where routine maintenance is problematic—for example, back-woods cabins or cottages that are only occasionally occupied. Sealed batteries are also used on sailboats and RVs where the rocking motion could spill the sulfuric acid contained in flooded lead-acid batteries. In these applications, space is limited and batteries are frequently crammed into out-of-the-way locations. Sealed batteries are also used in grid-tied systems with battery backup because these batteries tend to be ignored—out of sight, out of mind. However, if the batteries are deeply discharged for long periods, they can be ruined.

Sealed batteries offer several additional advantages over flooded lead-acid batteries. One advantage is that because they don't require periodic watering, they can be installed in tight quarters. Another advantage is that they charge faster. Sealed batteries also release no explosive gases, so there's no need to vent battery rooms or battery boxes where they're stored. In addition, sealed batteries are much more tolerant of low temperatures. They can even handle occasional freezing, although this is never recommended. Sealed batteries also experience a lower rate of self-discharge. That is, they discharge more slowly than flooded lead-acid batteries when not in use. (All batteries self-discharge when not in use.)

Sealed batteries do have a few significant disadvantages. One of them is that they are more expensive than flooded lead-acid batteries. Another is that they store less electricity than flooded lead-acid batteries of the same size. Sealed batteries have a shorter lifespan than flooded lead-acid batteries. In addition, they can't be rejuvenated (equalized) like a flooded lead-acid battery. Equalization is a controlled overcharge that produces a tremendous amount of hydrogen and oxygen gas. While safe in flooded lead-acid batteries (because the gases can escape through the caps) equalization of a sealed battery would result in excessive pressure buildup inside the battery. If they were accidentally equalized, pressure would be released through the pressure-release valve on the battery; if this happens, electrolyte is lost, which can destroy or seriously decrease the storage capacity of the battery. Finally, sealed batteries must be charged at a lower voltage setting than flooded lead-acid batteries. Although that's not a problem, you need to be sure to program your charge controller accordingly. Because of these problems, most installers working with battery-based systems usually recommend flooded lead-acid batteries.

Wiring a Battery Bank

Battery banks in wind systems are typically wired at 12, 24, or 48 volts. Twelve-volt systems are typically very small, for example, those used in sailboats and cabins. Although some 12-volt systems include inverters to boost the voltage to 120 volts, many of them run entirely off 12-volt DC electricity. To make this happen, you must install all 12-volt DC lights and appliances.

In larger systems, that is, those that power off-grid homes, batteries are typically wired to produce 24- or 48-volt DC electricity. The low-voltage DC electricity, however, is typically converted to 120 or 120/240 volt AC electricity.

Batteries in renewable energy systems are typically wired in series (positive to negative) and parallel. Each type of wiring achieves a different result. Let's begin with series wiring. Figure 8.6 shows four 6-volt batteries wired in series. Notice how the positive terminal (electrode) of one battery is wired to the negative terminal (electrode) of the next battery in series. Wiring four 6-volt batteries in series produces a 24-volt string. Also notice that each string has an unpaired positive and negative terminal and battery cable. These are commonly referred to as the "most positive" and "most negative." They run to the charge controller. Figure 8.6 also

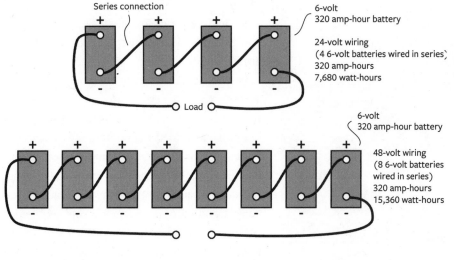

Series connection

+   +   +   +

6-volt
320 amp-hour battery

24-volt wiring
(4 6-volt batteries wired in series)
320 amp-hours
7,680 watt-hours

−   −   −   −

Load

6-volt
320 amp-hour battery

+   +   +   +   +   +   +   +

48-volt wiring
(8 6-volt batteries
wired in series)
320 amp-hours
15,360 watt-hours

−   −   −   −   −   −   −   −

FIGURE 8.6. Battery Wiring: 24- and 48-volt Systems. Batteries are wired in a string in series to boost the voltage. Several strings are wired in parallel. The battery banks here are wired for 24- and 48-volt electricity. Credit: Anil Rao.

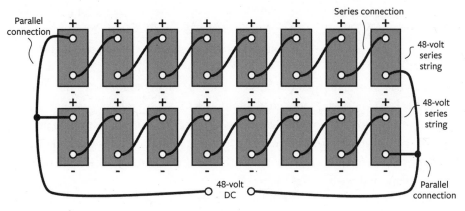

Parallel
connection

Series connection

+   +   +   +   +   +   +   +

48-volt
series
string

−   −   −   −   −   −   −   −

+   +   +   +   +   +   +   +

48-volt
series
string

−   −   −   −   −   −   −   −

48-volt
DC

Parallel
connection

FIGURE 8.7. Battery Wiring: Series and Parallel 24- and 48-volt Systems. In this diagram, two 48-volt strings are wired in parallel to boost the number of kilowatt-hours of electricity that are stored and available to a homeowner or business. Credit: Anil Rao.

shows a series string comprised of eight 6-volt batteries. Take a moment to study this figure.

While wiring in series boosts the voltage, wiring in parallel increases the amp-hour storage capacity of a battery bank—the number of amp-hours of electricity a battery bank can store. (One amp-hour is one amp of current flowing for one hour—either into or out of a battery bank.) To boost the amp-hour capacity of a renewable energy system, installers typically include two or more series strings of batteries in parallel, as shown in Figure 8.7. As you can see from the diagram, in parallel wiring the positive ends of each series string are connected, as are the negative ends. The most positive and most negative run to the charge controller and inverter.

ELECTRIC CARS like the author's Nissan Leaf are great commuter vehicles. They're cheaper to operate and maintain and when powered by solar are a truly sustainable form of transportation. Credit: Dan Chiras

Lithium Ion Batteries

Lead-acid batteries are the weak link in off-grid and grid-tied systems with battery backup. They are costly, don't store much electricity, and require proper housing (not too hot, not too cold). They also require periodic maintenance and more-frequent-than-you-will-be-happy-with replacement. This has led manufacturers to search for better options. Interestingly, much of the research and development on new batteries is occurring in the electric car industry. Companies such as Tesla, Nissan, and Ford—all leaders in this field—are intensely interested in developing inexpensive, long-lasting, lightweight, and high-capacity batteries to extend the range of their vehicles. In fact, today's electric cars, like my all-electric Nissan Leaf, are all powered by lithium ion batteries (Figure 8.8). Nissan's Leaf has a cruising range of 84 miles (if you purchase a vehicle with the 24-kWh battery) or 107 miles (with the 30-kWh battery). Both are perfect for most commuters. Tesla's Model S, a roomy sedan, travels about 270 miles on a single charge. How far you can go depends on how you drive, the type of road (gravel vs. pavement), road conditions (rain, snow, ice, or clear sailing), terrain (hills vs. flat), use of heaters or air conditioning, outdoor temperature, and the age of the battery.

Elon Musk, chairman of Tesla and First Solar, made headlines in 2015 when he announced a compact, wall-mounted lithium ion battery for household use, ostensibly as a backup for solar customers. My email inbox was soon flooded with questions from students, friends, and colleagues asking whether we've finally got a new battery to replace those clunky lead-acid batteries.

While exciting news, Tesla's Powerwall battery isn't quite what wind wind energy homeowners really need—at least not yet. Moreover, the inverter that comes with it is for solar electric systems.

At this writing (September 2016), Tesla currently offers two batteries, one with a 7-kWh capacity and another with 10-kWh capacity. For reference, a typical lead-acid battery bank in a 48-volt off-grid solar system would store over 50 kWh of electricity; however, to protect batteries and ensure long life, a homeowner shouldn't use more than half of that, so a typical battery bank in an off-grid home really only stores 25 kWh of electricity. To meet your needs, you'd need three Tesla Powerwalls for an energy-efficient off-grid system. (That's about a $10,000 investment at this writing.) But what about use in a home powered by a grid-tied system with battery backup?

Here the news is a bit more encouraging. In a system designed to meet critical loads—grid-tied with battery backup—a lead-acid battery bank would typically store about 18 kWh (eight 6-volt L16 lead-acid batteries that store about 360–370 amp-hours each). Remember, however, only half of this electricity is available for routine use. So, let's say these battery banks hold about 9 kWh of electricity.

In homes such as this, where the battery bank powers critical loads during a power outage, the 7-kWh and 10-kWh batteries might work. The 7-kWh battery, which is designed for daily use, would be a bit too small to power critical loads during a three- to four-day power outage. However, two or more Tesla batteries can be wired together to increase battery storage capacity. According to Tesla's website, the battery can produce 3,300 watts of continuous power. This would very likely not be sufficient for an all-electric home.

One additional consideration: Although the Tesla Powerwall battery looks sleek in photographs, it weighs 210 pounds (96 kg). It is also rather large, measuring 51.2″ × 33.9″ × 7.1″ (1300 mm × 860 mm × 180 mm)—that's over 4 feet by nearly 3 feet. Wall mounting would require a decent amount of space and special precautions to anchor the unit in place.

For most grid-tied solar systems with battery backup, Tesla's 10-kWh battery would be a better choice. According to the company's website, these batteries produce about 3,300 watts of continuous power. That would limit the number of loads that could be run at any one time. However, for a family that's tuned into energy savings, especially during times of power outages, this battery would probably work. But what would it cost?

In January 2016, the 7-kWh and 10-kWh Powerwalls were listed on Tesla Motors' website at $3,000 and $3,700, respectively, for installers. For comparison, a lead-acid battery bank with eight L16s would cost about $3,200 plus installation. Let's say $5,000. It would last 7 to 10 years. Powerwalls should last about the same length of time.

Pros and Cons of Lithium Ion Batteries

Manufacturers have developed over a half dozen types of lithium ion batteries, each with its own characteristics. One type is used strictly for consumer electronics. Others are being used in electric vehicles and solar batteries like the Tesla

Powerwall. One of the safest and most reliable is the lithium iron phosphate batteries. A more recent innovation, the lithium sulfur battery, appears to offer a significant advantage when it comes to performance-to-weight ratio. (That means it packs more punch per unit of weight.)

In general, lithium ion batteries can undergo many deep discharges. They also charge faster than lead-acid batteries because they can take the full charger output throughout the charge cycle. And, they don't suffer from sulfation, either. That said, lithium ion batteries require a battery charge management system that balances the charge on a per-cell basis and protects the battery from overheating. Lithium ion batteries also deteriorate over time. Many Nissan Leaf owners have found that the capacity of their battery banks declines each year. (It's a common complaint against what is otherwise a brilliantly designed and well-built vehicle.)

Lithium ion batteries may also require special care to ensure longevity. In their electric cars, for instance, Nissan recommends that the batteries not be charged to more than 80% of their full capacity to ensure longer life. If you need to charge up for a 78-mile trip, it's okay to charge the batteries to full capacity. Just don't do it every night, as owners are often wont to do. I set mine to charge to 80% most nights. The company also recommends charging using a 240-volt electric outlet, rather than 120. Rapid charging may shorten battery life, so be careful. Finally, for longer battery life, Nissan recommends letting the battery bank cool down a bit before plugging in the charger.

Lithium ion batteries, as a group, also don't function well in extreme temperatures, although the temperature range in which they function well is much wider than that of lead-acid batteries used in renewable energy systems. (That's why they work in cars in temperate climates.) Lithium ion batteries also perform less well when it's cold than during warmer weather, which is one reason why Nissan Leaf owners see their range drop over the winter. (The other reason is that snowy, icy roads create more friction that requires more energy to overcome. Running the electric heater also reduces range.)

So, while there are promising developments in the battery world, we are not quite there yet.

Sizing a Battery Bank

Properly sizing a battery bank is critical to designing a reliable off-grid wind system or grid-tied system with battery backup. The principal goal when sizing a battery bank for off-grid systems is to install a sufficient number of batteries to carry your household or business through periods when the wind is unavailable. These are called *battery days*, discussed shortly. In grid-tied systems with battery backup, the same goal applies; however, these battery banks are typically smaller because they only support critical loads.

In off-grid systems, the easiest way to size a battery bank is to calculate how much electricity you use in a day. Because electrical consumption varies from month to month, it is best to use a daily rate from the month with the greatest consumption. In many locations, this occurs during the dead of winter, the windiest time of year. People spend more time indoors. Lights are on more, as are TVs and computers. In hot, humid climates, however, peak demand may occur during the summer, when winds are weakest.

Once you determine how many kilowatt-hours of electricity you consume on average during a typical day during your most energy-intensive month, you must adjust for the number of windless days (battery days) that occur during that time. That is, you must determine how long your renewable energy system will be partially sidelined by low-to-nonexistent winds. In windier areas, such as many parts of Kansas, Oklahoma, and Texas, you may only need two or three battery days. In less-windy areas, like Missouri, you might require five battery days. Shorter periods are generally required for those who install hybrid systems—for example, wind and solar-electric systems. Fewer battery days may also be appropriate for people who run a backup fossil-fuel generator during cloudy periods. If you're interested in learning more about battery banks, read the accompanying sidebar, "More on Sizing a Battery Bank."

Reducing the Size of Battery Banks

Because batteries are expensive, require periodic maintenance, and take up a lot of room, many off-grid homesteaders install hybrid systems—often, these are systems that combine a PV system and a wind turbine. This allows them to reduce the size of their battery banks. Most homeowners add a gen-set—a gasoline, diesel, propane, or natural gas generator—to provide additional backup power. Or they install a larger wind turbine or a turbine on a taller tower. Because a wind turbine and tower—and PV modules—will likely outlast three to five battery banks, the investment in additional generating capacity may be well worth it in the long run. In addition, a PV system added to provide supplemental power, requires very little maintenance (except for batteries).

Home Power's Ian Woofenden says, however: "Although I love the theory of investing in RE capacity instead of lead (in batteries), there is a limit to its potential." In cloudy regions, PVs may not produce enough electricity to make a huge

More on Sizing a Battery Bank

To understand how a battery bank is sized, let's look at an example. Let's suppose that you and your off-grid family consume 4 kWh (4,000 watt-hours) of electricity a day on average during the most energy-intensive month of the year. If you need five days of battery backup, you will need 20 kWh of storage capacity (5 days × 4 kWh).

However, it's best not to discharge batteries below the 50% mark because deeply discharging batteries reduces their life. So, if you set 50% as your discharge goal, you'll need 40 kWh—or 40,000 watt-hours—of storage capacity.

To determine how many batteries you need to provide this amount of electricity, first find out how much electricity can be stored in the type of batteries you are using. Battery storage capacity is rated in amp-hours. For instance, a 6-volt L16 deep-cycle flooded lead-acid battery is rated at 360 amp-hours. But how do you know how many amp-hours of storage capacity you need when all you know is that you need 40 kWh of storage capacity?

The easiest way to figure this out is to convert amp-hours to kilowatt-hours. To make this conversion, simply multiply the voltage of the battery by the amp-storage capacity. This is a simple modification of the WAV equation introduced in Chapter 4: watts = amps × volts ($W = A \times V$). In this case, however, we want to determine watt-hour storage in each battery. The equation can be modified as follows: watt-hours = voltage × amp-hours. A 360 amp-hour 6-volt L16 battery, for example, stores 2,160 watt-hours or 2.16 kWh (when brand new).

The next task is to see how many batteries you will need to store this much electricity. To store 40 kWh of electricity, you'd need 18.5 batteries (40 kWh divided by 2.16 = 18.5 batteries).

If you wanted to wire your wind system at 48 volts, each string would have to contain eight 6-volt batteries. Two strings would contain 16 batteries. Since all series strings must be wired with the same number of batteries, you'd need to add a third string of eight batteries, or find a way to reduce your energy demand. If you chose the former, you'd end up with an off-grid system with 24 batteries consisting of three series strings, each with eight 6-volt batteries.

As already noted, battery storage can become quite expensive. At today's prices (September 2016), L16 batteries are running about $350 each. Unless you have an old set of batteries to trade in, you will very likely also have to pay a *core charge*, a fee that's tagged on to batteries if a buyer doesn't trade in an older battery at the time or purchase. (A core charge is very much like a deposit on a bottle or can. It's designed to encourage people to recycle their batteries.) That could come to $20 per battery. And, there may be shipping. All told, these batteries could easily cost $9,000 plus the costs of battery cables, a battery box, and installation.

difference during winter months. In such instances, batteries and gen-sets may be a preferred option. Ian should know. He lives in the cloudy Pacific Northwest.

Fortunately, the National Renewable Energy Laboratory (NREL) offers online assistance to people trying to figure out the best combinations. The program, called *HOMER*, is designed to optimize hybrid power systems. "This is a great tool for optimizing the size of various components in a hybrid power system," notes NREL's small-wind expert Jim Green. "It typically shows that adding a backup generator will reduce system cost and that optimum battery bank size may be on the order of 8 to 12 hours of storage, much less than 3 days, which is typically used." He adds: "Of course, the optimization point will change over time as the price of propane and natural gas go up. Even so, the generator run time can be quite low.... This model [HOMER] will be more than most homeowners will want to tackle, but some will be able and willing to use it. Equipment dealers find it to be a useful tool."

Battery Maintenance and Safety

Now that you understand lead-acid batteries, your options, and a little about wiring and sizing a battery bank, it is time to turn to the equally important topics of battery care, maintenance, and safety. Battery care and maintenance are vital to the long-term success of a renewable energy system. Proper maintenance increases the service life of a battery. Because batteries are expensive, longer service life results in lower operating costs and less environmental impact over the long haul. Put another way, the longer your batteries last, the cheaper your electricity will be and the less impact you will have on the air, water, and soil. In this section, we'll explore battery care, maintenance, and some safety issues.

Keep Them Warm

Batteries may be the workhorses of a renewable energy system, but they like to be kept warm—and full. Cold conditions dramatically reduce their capacity— the amount of electricity they'll store (Figure 8.9). That's because low temperatures slow down the chemical reactions in batteries, reducing electrical storage.

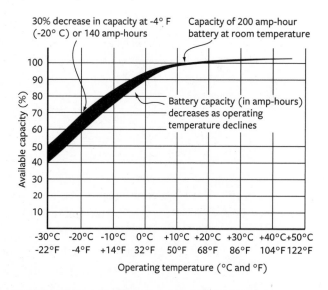

FIGURE 8.9. Battery Performance at Various Temperatures. Batteries function optimally above 50°F. Battery capacity decreases rather dramatically as temperatures in the battery room fall. Credit: Anil Rao.

While low temperatures reduce battery efficiency, higher temperatures result in an increase in outgassing. Outgassing is the release of hydrogen and oxygen gas. It occurs primarily as a result of electrolysis, as already noted. Outgassing not only produces explosive gases, it also reduces battery fluid level. The main danger is permitting the water level to drop below the top of the lead plates. This will quickly kill a battery. Higher temperatures also lead to higher rates of self-discharge. Outgassing is not a trivial matter. Despite what you might think, Arizona is a harsher climate for car batteries than Minnesota.

For optimal function, batteries should be kept at around 75°F to 80°F (24°C to 27°C)—about the same temperatures that we like. In this range, batteries will store and hence release a lot more electricity. Guaranteed!

If you can't ensure this narrow temperature range, shoot for a range between 50°F and 80°F. Rarely should batteries fall below 40°F or exceed 100°F—that's 4.5°C and 38°C.

Ideally, batteries should be stored in a separate battery room or a battery box inside a conditioned (heated and cooled) space held at the optimum temperature. Basements make good candidates for battery rooms or battery boxes. An unconditioned basement (not heated or cooled) tends to range between 60°F (16°C) in the winter and 70°F (21°C) in the summer. (Be sure to check the basement temperature in your home; don't take my word for it. Yours might get hotter or colder.) Modestly heated and cooled shops or utility rooms may also work. I discourage people from housing batteries inside their homes, unless they can build a separate room. I housed mine in Colorado in a well-sealed, ventilated, specially-built closet in my utility room.

Whatever you do, don't store batteries in cold buildings—for example, garages, barns, or sheds. Expect very short service lives when batteries are kept in cold environments. Furthermore, when batteries are in a low state of charge, they can freeze more easily. Freezing causes the water inside a battery to expand and can crack the plastic cases, allowing sulfuric acid to leak out, creating a dangerous mess in the battery room. (Deeply discharged batteries are especially prone to freezing because the sulfuric acid has reacted with lead and lead dioxide to form lead sulfate, resulting in a higher concentration of water, which makes the battery more prone to freezing.) As just noted, extremely hot environments are also not recommended.

Batteries should not be stored on concrete floors. Cold floors cool them down and reduce their capacity. Always raise batteries off the floor, so they stay warmer.

If you must store batteries outside your home, be sure to heat and cool the building. Heating and cooling a shed, garage, or barn can be quite expensive and waste valuable fossil-fuel energy. Plus, it can gobble up most, if not all, of the energy a renewable energy system generates, perhaps even more. In 2015, I troubleshot a grid-tied system with battery backup installed by an installer with little experience in battery-based systems. The homeowners were upset because their winter electric bill was higher *after* they installed their PV system. They assumed that the PV system was not working at all.

When I visited the site, I found out that they installed a portable electric heater in their detached garage where the batteries were housed, and they ran it all winter long. They did this on the advice of the installer. He'd told them they'd need to keep the batteries warm for optimum performance. Unfortunately, electric heaters consume tons of electricity—more than a medium-sized solar system can produce. The same could occur with a wind energy system.

If you must heat a battery room, insulate it and retrofit it for passive solar by installing south-facing windows. Be sure to install window shades that can be closed at night to hold heat in the building. Insulation will help keep the building warmer in the winter and cooler in the summer. Passive solar design will allow the low-angled winter sun to warm the building. A solar hot air collector might also help keep a garage or shed warmer. (For more on passive solar design, see my book *The Solar House: Passive Heating and Cooling*. For information on solar hot air collectors, check out my book *The Homeowner's Guide to Renewable Energy*.)

Ventilating Flooded Lead-Acid Batteries

To ensure safe operation, battery boxes must be ventilated to the outside to remove potentially explosive hydrogen and oxygen gases released when flooded batteries charge. (This does not apply to sealed batteries.) A small 2-inch (5-cm) vent made from PVC or ABS pipe is all that's typically required to vent a battery box. The outlet should be placed high, as hydrogen is fairly light and rises to the top of the box.

Battery rooms may also require venting (Figure 8.10). However, hydrogen disperses fairly rapidly in a medium-to-large room and requires very little air movement to prevent it from building up to dangerous levels. Natural air movement in a battery room combined with normal air changes that occur in occupied spaces—or as a result of heating systems—are typically sufficient to disperse oxygen and hydrogen gases generated in most battery rooms. Remember: It all depends on the size of the room. The larger the room, the less likely venting will be required. When in doubt, err on the conservative side. Ventilate.

When venting a battery box or a battery room, be sure that the vent system doesn't cool the space in the winter. Place the opening on the downwind side of

FIGURE 8.10. Battery Vent System. Flooded lead-acid batteries require venting to remove hydrogen released when batteries are charging from a wind turbine, gen-set, or grid. Vents drawn here are supersized to emphasize their importance. In most cases, smaller passive vents are required. Be sure to allow replacement air to enter near the bottom of the battery room. Credit: Anil Rao.

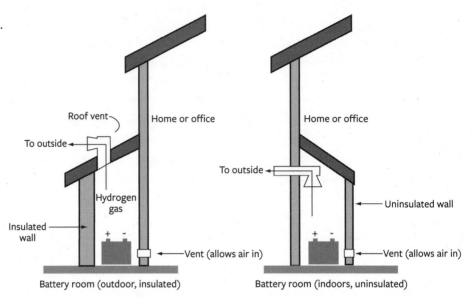

the building. Also, be sure that appliances and devices that require an open flame (a gas water heater, for example) or that might produce sparks (an arc welder, for example) are not housed near a battery bank.

As illustrated in Figure 8.10, proper ventilation requires an air outlet near the top of the ceiling in a battery room (or battery box) to vent hydrogen gas to the outside. It also requires small air inlets near floor level. (When ventilating a battery box, don't penetrate the waterproof lining that is installed to capture electrolyte should a battery leak.) These openings allow fresh air into the battery room or battery box. Check with your building department to be sure you comply with their requirements, if any.

Battery rooms are generally passively ventilated. However, they can also be actively vented or power vented—for example, if the pipe run is long and/or contorted (not a straight run to the outside). Power venting requires a small electric fan that exhausts hydrogen gas while batteries are charging.

As nifty as power venting may seem, it's generally not necessary. Hydrogen is extremely light and easily escapes if a room is properly vented. So unless the vent pipe is long and contorted or you live in a cold climate, you probably don't need to power vent a battery bank. If, however, batteries are located in a room used for other purposes, for example, a shop or garage, active venting may be prudent. It reduces the chance of explosions.

Battery Boxes

Battery boxes are typically built from plywood (Figure 8.11). Some individuals build them with wooden lids; others incorporate clear plastic (polycarbonate) in their lids. Be sure to seal the lids to prevent hydrogen from escaping. And, be sure to line the boxes with an acid-resistant liner to contain possible acid spills from cracked battery cases. I use black ABS plastic to line battery boxes. I purchase it in 4 × 8 sheets, cut it with a circular saw (fine-tooth saw blade), and glue it in place with silicone caulk. I then caulk all the joints. Don't use polyethylene plastic. Although it is cheaper than ABS, it won't hold silicone caulk. As an alternative, battery leakage can be prevented by installing batteries in durable plastic tubs inside the battery box (Figure 8.12). Choose heavy-duty polyethylene tubs.

Lids on wooden battery boxes should be hinged and sloped to discourage people from piling items on them. The steeper the lid, the better.

As noted above, be sure to place battery boxes in a warm room and always raise boxes off cold concrete floors. I install 2 × 4s under my battery boxes to keep them off concrete floors. I've even insulated the floors

FIGURE 8.12. Battery Boots and Plastic Tubs. I installed individual battery boots on the batteries in my off-grid home in Colorado to contain possible leaks. Durable plastic tubs can also be used. Don't buy cheap ones. Batteries are heavy and could cause tubs to crack. Credit: Dan Chiras.

FIGURE 8.11. Battery Box. We made this battery box from plywood. Be sure to vent and seal the battery boxes when installing flooded lead-acid batteries just in case a battery leaks. This photo shows the ABS plastic lining we caulked in place to seal the battery box to prevent potential leakage. Credit: Dan Chiras.

of some battery boxes to reduce heat loss. Also be sure that the boxes or shelves you build are strong enough to support the weight of the batteries. Large RE batteries weigh 80 to 125 pounds each—that's 36 to 57 kg. Also, be sure to install a lock on battery boxes to keep kids out.

Battery storage boxes can be purchased from commercial outlets. Radiant Solar Tech, at radiantsolartech.com, for example, manufactures sealed plastic battery boxes with removable lids. This company will also custom build boxes for homeowners.

When shopping for battery boxes, note that many commercially available products are designed for sealed batteries, not flooded lead-acid batteries. They are often metal cabinets. So, before you buy a manufactured battery storage cabinet, be sure that it will work with the type of batteries you are planning on using in your system. Cabinets for flooded lead-acid battery banks should provide sufficient head clearance so you can view electrolyte levels, and there has to be enough room to allow you to fill each cell with distilled water. Because sealed batteries don't require watering, the shelves are usually spaced very close together, which would make it impossible to service flooded lead-acid batteries—that is, check electrolyte levels and fill the batteries with distilled water.

Finally, batteries should be located as close to the inverter and charge controller as possible. These wire runs require very large diameter battery wire, which is expensive. Minimizing the distance saves money, and it minimizes voltage drop, which will help make your system more efficient.

Keep Kids Out

I mentioned it earlier, but it bears repeating: battery rooms and battery boxes should be locked—*especially* if young children are likely to be about. This will prevent kids from coming in contact with the batteries and receiving electrical shocks that could be fatal or potentially disfiguring acid burns.

Managing Charge to Ensure Longer Battery Life

Housing flooded lead-acid batteries at the proper temperature and keeping them topped off with distilled water ensures a long life. Longevity can also be ensured by proper charge management—keeping batteries as fully charged as possible. Battery banks should also be equipped with temperature sensors that connect to battery chargers. They help regulate charging according to room temperature.

To understand why it is so important to manage battery charge, let's take a look at *cell capacity*, the amount of electricity a battery can store.

"Cell capacity is the total amount of electricity that can be drawn from a fully charged battery until it is discharged to a specified battery voltage," according to Richard J. Komp, author of *Practical Photovoltaics: Electricity from Solar Cells*. Battery cell capacity is measured in amp-hours. A Trojan L16H battery (H stands for high-capacity) can store 420 amp-hours of electricity. Theoretically, a battery with a 420-amp-hour storage capacity could deliver one amp of electricity over a 420-hour-period, or 420 amps for one hour. In reality, the amount of electricity that can be drawn from a battery varies with the discharge rate, that is, how fast a battery is discharged. As a rule, the faster a lead-acid battery is discharged, the less you'll get out of it. A lead-acid battery discharged over a 20-hour period, for example, will yield 100% of its rated capacity. Discharge that battery in an hour and a half, and it will deliver only 75% of its rated capacity. Figure 8.13 illustrates this concept.

To standardize the industry, most batteries are rated at a specific discharge rate, usually 20 hours (which is what the 420-amp-hour capacity in the example above is based on).

Like many devices, lead-acid batteries last longer the less you use them. More specifically, the fewer times a battery is deeply discharged, the longer it will last. As illustrated in Figure 8.14, a lead-acid battery that's regularly discharged to 50% will last for slightly more than 600 cycles, if recharged after each deep discharge. If regularly discharged

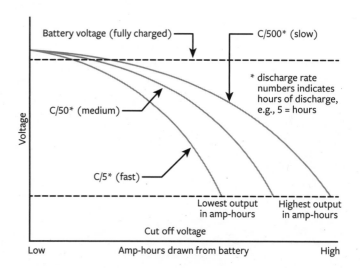

FIGURE 8.13. Discharge Rate vs. Amp-hours Delivered. The rate at which a battery bank is discharged determines how much energy can be obtained. The slower the rate of discharge, the more energy. Credit: Anil Rao.

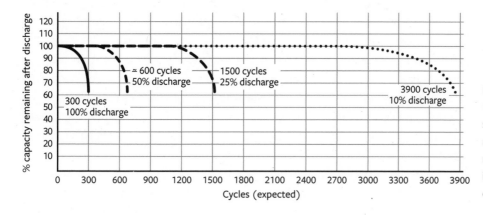

FIGURE 8.14. Battery Life vs. Deep Discharge. This graph shows that shallow discharging results in more cycles. As a result, batteries will last longer if they are not discharged as deeply. Credit: Anil Rao.

no more than 25% of its rated capacity—and recharged after each discharge—the battery should last about 1,500 cycles. If the battery is regularly discharged only 10% of its capacity, it will last for 3,600 cycles. So, even though deep-cycle batteries can handle deep discharges, shallow cycling makes them last longer.

As renewable energy expert Ian Woofenden points out, however, this topic—like so many issues in life—is a bit more complicated. While deep discharging reduces the lifespan of a battery, renewable energy users are more concerned with the cost of the battery per watt-hour cycled. In other words, what we want from batteries is not simply to "last a long time" but to cycle a lot of energy at an affordable price.

Studying the cost of a battery per cycled kWh, says Ian, provides a better measure than total years of service. Theoretically, you'll get the most bang for your buck by cycling in the 40% to 60% discharge range. This goes against the "shallow cycling makes batteries last longer" idea. (After all, if you double the size of a battery bank and reduce the discharge cycle by half, you'd only break even on the larger investment.)

Battery life also depends on frequent recharging of batteries after deep discharges. As you may recall, charging removes lead sulfate from the lead plates. Whatever you do, never leave batteries at a low state of charge for a long time. Lead sulfate crystals will enlarge and could damage your batteries. A good rule to follow is to be sure that the battery bank is fully charged at least once a week (and fully charged does not mean mostly charged!). Think of it this way, if you don't top them off, you start losing the top, starting a downhill slide.

Protecting batteries from deep discharge is easier said than done. As Ian points out, "in real life, you often don't have such close control over how deeply your batteries cycle." If your system is small, and you don't pay much attention to electrical use, you'll very likely overshoot the 40% to 60% mark time and time again. If you monitor your electrical usage more carefully, and charge batteries shortly after periods of deep discharge, you are more likely to achieve your goal.

One way of reducing deep discharge is to conserve energy and use electricity as efficiently as possible. Conserving energy means not leaving lights or electronic devices running when they're not in use—all the stuff your parents told you when you were a kid. Energy efficiency means installing energy-efficient lighting, appliances, electronics, and so on—the ideas energy conservation experts have been suggesting for decades. In addition, it means eliminating phantom loads.

Energy conservation and energy efficiency are only half the battle, however. In an off-grid system, you will also have to adjust electrical use according to the state of charge of your batteries. In other words, you will need to cut back on electrical usage when batteries are more deeply discharged and shift your use of electricity to windy days when the batteries are more fully charged. You may, for instance, run your washing machine and microwave on a windy day when your batteries are full, but hold off during windless periods when batteries are running low—unless you want to run a backup generator. Learning how to monitor the state of charge in your batteries is a skill you can learn. It requires a high degree of awareness of what's going on outside, however.

Experienced battery operators watch the weather and their battery banks like hawks. If they experience a couple of windless days, they know that they'll have to cut back on electrical consumption, for example, by cooking on a woodstove or a gas-powered range, as opposed to a microwave.

I monitored my off-grid solar and wind system in Colorado by checking the battery voltage on the LCD screen on my system at the *end* of each day, when the batteries were no longer charging and demand for power in the house was minimal. The voltage reading gave me a good idea about the state of charge of my batteries. The higher the voltage, the more electricity my batteries held.

Voltage is a surrogate measure of state of charge. That is, it is a good approximation of battery capacity. Remember, however, battery voltage and state of charge are not directly related unless the batteries are at rest. At rest means they are not being charged or discharged or have not been recently charged or discharged. That's because the voltage of batteries that are being charged will read a bit higher than when charging ceases. Over the years, I've found that it doesn't take very long for a battery voltage to settle down once charging ends.

A more precise way to monitor the state of batteries is with a digital amp-hour or watt-hour meter. One popular meter, the TriMetric, keeps track of the number of amp-hours of electricity moving in and out of a battery continuously. By doing so, and by it knowing the size of the battery in amp-hours, which is programmed into the meter, it can tell you how many amp-hours are stored in the battery at any point in time. To make your life easier, the TriMetric displays a "fuel gauge" reading that shows the state of charge as a percentage, similar to those on laptop computers. Similar meters can be ordered with charge controllers and inverters for battery-based systems.

Homeowners should use battery state-of-charge information to adjust daily activities. If batteries are approaching the 40% to 60% discharge mark, hold off on activities that consume lots of electricity—like vacuuming, toasting bread, or running a hair dryer.

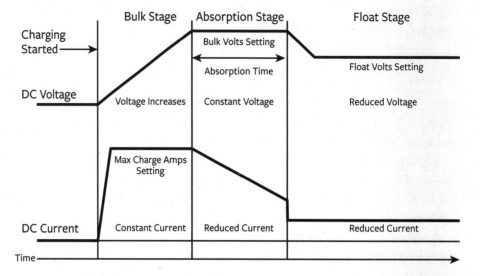

FIGURE 8.15. Battery Charging. Charge controllers and battery chargers in battery-based RE systems carefully control the flow of electricity into a battery bank. For a description, see text. Credit: Apollo Solar.

Another way to minimize deep discharging is to oversize a wind system—that is, to install a larger turbine or a turbine on a taller tower (the latter is generally more economical). Or, you can install a hybrid system—for example, add a PV array to supplement your wind turbine. This will ensure that your batteries are supplied with more electricity throughout the day and will lessen chances of deep discharging. Oversizing a battery bank, but not the power source has just the opposite effect. Batteries may remain in lower states of charge for longer periods.

A backup generator can also be used to manage batteries. During long windless periods, for instance, a homeowner can fire up a generator to boost battery voltage. Generators, or gen-sets, as they are also called, can be wired directly to the inverter in a battery-based system so they start automatically when the battery voltage drops to a predetermined level. Bear in mind that not all inverters and not all generators operate automatically, so if you want this feature, be sure your generator and inverter are designed for automatic operation. For more on battery charging, see the accompanying textbox "Understanding Battery Charging."

Watering and Cleaning Batteries

As noted earlier, the lead plates in batteries are immersed in a 30% solution of sulfuric acid. Sulfuric acid participates in reversible chemical reactions that store and release electrons. Because these reactions are reversible, you'd think that battery fluid levels would remain constant over the long haul. Unfortunately, that's not the case. Battery fluid levels decrease over time for three reasons. The first,

Understanding Battery Charging

Batteries are charged by the wind turbine via the charge controller when the wind turbine is spinning. The charge controller performs several key functions. It feeds DC electricity from the turbine to the battery bank in a highly controlled manner. It also prevents batteries from overcharging. It does this by terminating the flow of DC electricity when the battery bank is full—voltage only climbs to a programmed set point.

As you may recall from Chapter 7, batteries are also charged by the battery charger in battery-based inverters. The charger converts 120-volt AC electricity from the gen-set to DC to charge batteries. In grid-connected systems with battery backup, they can convert AC from the grid to DC electricity. Converting AC to DC is the function of a rectifier. Battery chargers produce DC electricity at the same voltage of the battery bank so it matches the voltage at which it was wired.

Battery charging by the charge controller and battery charger occurs in three stages—if the batteries have been deeply discharged. These stages are controlled by the charge controller, if the battery bank is being charged by the wind turbine. If the battery bank is being charged by the battery charger in the inverter, that controls the charging.

The first stage of battery charging is known as the *bulk stage*. During this stage, shown on the left in Figure 8.15, the charge controller (or inverter-based battery charger) delivers a constant and relatively high charge (lots of amps). The batteries are forced fed, bulked up, like a runner preparing for a marathon.

As the charging source "pumps" amps into the battery bank, the battery voltage steadily increases. Once the battery voltage reaches a certain point, known as the *voltage regulation set point*, or *bulk volts setting*, the bulk stage ends. At the end of the bulk stage, the battery is about 80%–90% charged.

In the second stage, known as the *absorption stage*, the battery voltage is kept constant and the charge rate (amps) slowly declines. The absorption stage typically lasts about two hours to ensure that a battery is fully charged. Absorption time is determined by the type of battery and is an adjustable setting. When an installer first programs a battery-based inverter, he or she is asked the type of battery, and the inverter then sets the duration of the absorption stage.

At the end of the absorption stage, the charging cycle enters the third and final stage, the *float stage*. This occurs if the batteries are not being used. During the float stage, the voltage is reduced slightly and held constant at the float volt setting. The current is maintained at a level just a little higher than that which is needed to offset the self-discharge rate of the battery. The float stage is akin to trickle charging a motorcycle or lawn tractor battery over the winter to prevent self-discharge that could, over the course of a winter, run a battery dry.

In an off-grid system in which the generator is controlled automatically by the inverter, the generator is switched off as soon as the charger enters the float stage. This minimizes generator run time and fuel consumption. In a grid-connected system, the float stage may continue indefinitely with current supplied by the charge controller via the wind turbine or the inverter-based battery charger when charged via the utility grid. Many inverters in grid-tied systems with battery backup, however, shut down when not being charged by a renewable energy source to reduce dependence on the grid. (Any self-discharge that occurs during this period is then made up by the wind generator when it starts spinning again.)

and most important, is hydrolysis, the breakdown of water by electricity. Remember that water splits into oxygen and hydrogen, both of which can escape from a flooded lead-acid battery. Second, a small amount of water can evaporate and escape through battery cap vents. Third, sulfuric acid can escape through the vents during charging. It escapes as a result of the strong bubbling action of hydrogen and oxygen gases that occurs during charging. This results in the production of a mist of acid that precipitates out on the top of the battery. When it dries, it leaves behind a fine powder of sulfuric acid.

All of these potential sources of battery fluid loss add up over time and can run a battery dry. When the plates are exposed to air, they quickly begin to corrode. A battery's life is pretty well over when this happens. I have tried all kinds of tricks to bring batteries that have fallen victim to low-electrolyte levels back to life, but to no avail. Even if only one cell in a battery goes bad, the entire battery is out of commission. If a battery bank is more than a year to a year and a half old, you can't replace a bad battery in the system with a new one. They generally all have to go—or, at least that's the advice of battery experts. (You could probably replace the bad battery with a brand new battery, but it will operate at the same level as the remaining batteries, which will be lower.)

To prevent this potentially costly occurrence, monitor battery fluid levels regularly. Many experts recommend checking batteries every month. Others recommend checking batteries every three months. I have found that a monthly checkup works best for my off-grid systems, and two- to three-month checkups work best for my grid-tied system with battery backup. Be sure to note battery checkups on your calendar or enter them into your daily reminders on your cell phone. Battery watering is very easy to forget.

Be careful not to get lulled into complacency after installing new batteries, however. The first year of a battery's existence is much like the year or two after a honeymoon. Peace and happiness prevail. That is to say, brand new batteries operate very smoothly for a year or so with very little water loss. As time passes, however, batteries, like our partners in marriage, can become more demanding—a lot more demanding. They require more frequent inspection and watering. And remember, divorce from your battery bank is not an option if you are living off grid. Remember, too, that batteries in hotter climates will require more frequent watering than batteries in cooler climates. In the Dominican Republic, for instance, you might need to water batteries twice a month during the sunniest part of the year. Systems that experience high discharging and charging rates will also require more frequent watering because of frequent hydrolysis.

To check fluid levels, unscrew the caps and peer into each cell when the system is not charging. Use a flashlight, if necessary—*never a flame* from a cigarette lighter or any other source! Be sure that the batteries are installed in their battery room or battery box in a way that allows them to be easily inspected. Avoid stacking batteries on shelves that preclude you from safely peering into each of the cells. That means shelves need to be widely spaced. You'll also need room to fill the batteries, a topic discussed below.

Battery acid should cover the plates at all times—at a bare minimum at least one-fourth of an inch above the plates. (That's usually about ¼ inch below the vent or fill tube—the opening in the battery casing into which the battery cap is screwed.) It's better to keep the battery more fully filled, but don't overfill it, either. Over-filling a battery will result in battery acid bubbling out of the cells when batteries are charged. As you just learned, water evaporating from the acid on the top of the battery leaves a white acidic deposit. It not only looks messy, it supposedly can conduct electricity, slowly draining the batteries. (Though I have trouble believing that.) Battery acid also corrodes metal—electrical connections, battery terminals, and battery cables. In addition, the loss of battery acid will result in a dilution of the electrolyte within the battery cells because the addition of distilled water to compensate for lost fluid reduces the sulfuric acid concentration.

When filling a battery, be sure to add only distilled (or deionized) water. Never use tap water. It may contain minerals or chemicals that could contaminate the battery fluid and reduce a battery's lifespan.

Batteries should be 100% charged—or close to it—before filling them with distilled water. This occurs on windy days when demand is low or nonexistent. You can also fill batteries right after equalization.

Don't fill depleted batteries and then charge them with a backup generator. This could cause battery fluid to bubble out. If the fluid levels are dangerously low, however, add a tiny bit of distilled water, then fully charge the batteries. After that, it is safe to top off the batteries with distilled water.

Battery acid bubbling out of batteries should be cleaned promptly, although most of us don't monitor our batteries that closely. Use paper towels or a clean rag to sop up the liquid. If you don't monitor your batteries to detect spills, be sure to check for white acid powder each time you fill your batteries. Some sources recommend using a solution of baking soda (sodium bicarbonate) to neutralize battery acid that bubbles up onto the top of batteries. After carefully rinsing batteries with sodium bicarbonate, they say, batteries should be wiped down with distilled water again. I discourage people from cleaning batteries with baking soda because it could seep into the cells of a battery, neutralizing the acid and reducing battery capacity. It's been my experience that a damp paper towel or rag is sufficient to clean the surface of batteries. When cleaning batteries, be sure to wear gloves, protective eyewear, and a long-sleeved shirt—one you don't care about. If you get acid on your skin, wash it off immediately with soap and water. If you feel a burning itch on exposed skin, it means you've got a little acid on you. Wash it off immediately and the burning should go away.

Although I don't recommend using baking soda to clean batteries, it is a good idea to keep a few boxes of baking soda near your batteries in case there's an acid spill in the battery room. Acid spills are extremely rare but may occur if a battery case cracks. If one occurs, neutralize the acid with baking soda before mopping it up. You should also install a dry chemical fire extinguisher for safety.

When filling batteries, be sure to take off watches, rings, and other jewelry, especially loose-fitting jewelry. Also empty shirt pockets of cell phones, metal-frame eye glasses, pens, and such. Metal objects such as jewelry will conduct electricity if they contact the positive and negative battery terminals. Such an event will convert your jewelry to a puddle of metal and could cause a very serious burn. One 6-volt battery could produce over 3,000 amps if the positive and negative terminals of a battery are shorted. As if that's not enough, sparks could ignite hydrogen and oxygen gas in the vicinity. Heat produced by a short circuit could cause the battery case to explode, ejecting hot battery acid in all directions. You'll surely lose an eye or two if you are not careful. One of my students used to work for an automobile manufacturer and lost an eye when a battery exploded on him.

Also, be careful with tools when working on batteries—for example, when tightening cable connections. A metal tool that makes a connection between oppositely charged terminals on a battery becomes red hot and may be instantaneously welded in place. It can also ignite hydrogen gas, causing an explosion. Short-circuiting a battery will also very likely ruin the battery. To prevent these problems, be sure to wrap hand tools used for battery maintenance in electrical tape so that only one inch of metal is exposed on the working end (that way, it can't make an electrical connection). Or, buy insulated tools to prevent this from happening. I use a socket wrench with a stout rubber handle. (It saved my life on one occasion when I was working on wiring at the inverter.) It is a good idea to have a dedicated set of tools (box wrench or socket wrench) for use only on the battery bank. Find a place for them in a box next to the batteries. That way, you won't be tempted to use whatever (non-insulated) tools might be at hand at the moment you need them.

Also when working on batteries, for example, when wiring them for the first time, be sure never to place a tool on a battery or the battery box. Place them on the ground next to you or some other secure place where they won't accidentally be pushed against the battery posts.

You may need to clean the battery terminals every year or two. As noted earlier, the fine corrosive mist containing sulfuric acid released when a battery charges corrodes the metal posts, battery cables, and hardware. It forms a white/turquoise gooey powder. (The green is from the corrosion of the copper wires in battery cables.) Corrosion increases resistance in the circuit that will reduce battery performance (efficiency) when charging and discharging. Even if corrosion only occurs on one battery terminal, every battery in the series string will be affected. Corrosion is often indicative of a loose connection.

To clean the posts, shut down the system. Disconnect the battery bank from the charge controller. Remove the battery cables that need cleaning. Then use a small wire brush and, if necessary, a spray-on battery cleaner. Cleaners can be purchased at auto parts stores or well-equipped hardware stores. Before you reconnect the batteries, be sure to wire-brush the battery terminals so they are shiny. After they are cleaned, reinstall the nuts and bolts, or replace them with new ones. Tighten bolts carefully, but completely. No loose connections! (They can accelerate corrosion.)

To reduce maintenance, coat battery posts and the ends of the battery cables with a battery protector/sealer, available at auto supply stores and some hardware stores. It's sprayed on the ends of the battery cables and terminals where it forms a protective layer that halts or greatly reduces corrosion. (Some homeowners coat battery terminals with petroleum jelly. Unfortunately, this product may collect dust over time, and it is messy to work with when you need to remove a battery cable or tighten a bolt.) Be sure to use flat washers and lock nuts when installing battery cables.

I sometimes use flat copper bar stock instead of battery cables. This type of battery connection is easier to make than battery cables, which require you to carefully crimp a terminal on each end. To make a connector with flat copper bar stock, simply cut the material to length and then drill an appropriately sized hole on each end. Before you cut it to length, however, be sure to see if you will need to bend the stock to fit. If so, be sure to cut a bit longer. I have found that copper bar stays very clean and presents a lot less corrosion than standard battery cables. I use ⅛-inch thick, ¾-inch wide copper bar stock—it's called Multipurpose 110 copper. I purchase it from McMaster-Carr, one of my favorite online sources of hardware, fans, and just about everything else I need. Their delivery is mind boggling. Order one afternoon from Chicago, and the product is in my hands the next day!

Equalization

Another key to ensuring the longest possible battery life is periodic equalization. *Equalization* is a carefully controlled overcharge of batteries that, among other things, removes lead sulfate from lead plates. If done properly, meaning at the right times of the year, periodic equalization can greatly extend battery life.

Batteries can be equalized in several ways. In some cases, they can be equalized by a wind turbine or a PV array in a hybrid system. A strong storm in the winter or early spring, for example, can provide enough electricity to equalize a battery bank using a wind turbine. In Colorado, I equalized batteries in my off-grid wind and solar system during the summer using the PV array. It's possible to do this only if it's sunny and the batteries are pretty full. (Unfortunately, summertime equalization is not often needed.)

Equalization is most commonly carried out by running a backup generator. Periodic equalization is performed for three reasons: The first is to drive lead sulfate crystals off the lead plates, preventing the formation of larger crystals, which reduce battery capacity, are difficult to remove, and therefore can permanently damage the lead plates.

Sulfate and Sulfation, What's Normal and What's Bad

During discharge in a lead-acid battery, lead sulfate, $PbSO_4$, is formed on the positive and negative plates. When first produced, lead sulfate forms a soft deposit on the plates. If the battery is promptly and fully recharged, lead sulfate is converted back into plate material (lead on the negative plates and lead dioxide on the positive plates) and sulfuric acid (electrolyte). However, if the battery is left discharged, even partially discharged, for some time, the lead sulfate slowly forms large crystals. These crystals are hard and are not easily converted back into plate material during a subsequent charge. An accumulation of these hard crystals is called *sulfation*.

Sulfation is bad for a couple reasons. First, lead sulfate is an insulator. It accumulates between the electrolyte and the lead plates, thus reducing the chemical reactivity of the plates. Lead sulfate crystals therefore effectively reduce the surface area of the plates. In effect, batteries get smaller and their capacity to store energy is reduced. Second, hard crystals of lead sulfate eventually break off and fall to the bottom of the cell, reducing plate material. Not only is useful plate material lost forever, but the accumulation in the bottom of the cell eventually reaches the bottom of the plates. This prevents electrolyte from contacting the surface of the plates. At this stage, if not before, the battery is so degraded that it must be replaced. Lead accumulating in the bottom of each cell can also short out plates, ruining a battery.

Batteries are also periodically equalized to stir up (de-stratify) the electrolyte. Sulfuric acid is heavier than water and tends to sink, settling near the bottom of the cells in flooded lead-acid batteries. During equalization, hydrogen and oxygen gases released by the breakdown of water create bubbles that cause the battery to "boil." (This phenomenon is called *boiling* because the rapid formation of bubbles resembles boiling in water; it is not caused by high temperatures.) The bubbles mix the fluid so that the acid is de-stratified in each cell of each battery, which improves efficiency.

Equalization is also performed to help bring all of the cells in a battery bank to the same voltage. That is, it equalizes voltage in the cells of a battery bank. That's important—indeed vital—to a battery bank because some cells may sulfate more than others. As a result, their voltage may be lower than others. A single low-voltage cell in one battery reduces the voltage of the entire string. Batteries and battery banks operate much like a camel train. A camel train operates at the speed of its slowest camel. Similarly, a battery operates at the voltage of its lowest cell. Low voltage in one cell drags the voltage of the entire battery down. One low battery, in turn, drags the voltage of the entire battery bank down.

If equalization restores batteries, why won't a battery remain functional for an eternity? Although equalization removes lead sulfate crystals from plates, restoring their function, some lead sulfate crystals are dislodged—flake off—during equalization, settling to the bottom of the batteries. As a result, batteries lose lead over time and never regain their full capacity. This sloughing slowly whittles away at the lead plates. As the amount of lead decreases, the battery's ability to store electricity declines. Lead that sloughs off positive and negative plates also builds up on the bottom of each cell, and can short out a cell.

Equalization is fairly simple. In systems with gen-sets, the owner simply sets the inverter to the equalization mode and then cranks up the generator. The inverter or charge controller controls the process from that point onward. In wind/PV hybrid systems, the operator sets the wind generator charge controller to the equalize setting during a period of high wind. The controller takes over from there. Or, he or she can use the solar system to equalize during extremely sunny periods.

During equalization, the charge current (number of amps fed into a battery bank) is kept relatively constant and high for a set period, usually four to six hours. During this process, the battery voltage is allowed to rise to higher than normal levels. (Normally the charge controller shuts off or slows the flow of electricity to batteries once a certain voltage is reached.) Over time, the voltage of all of the cells of all the batteries in a battery bank rises to the same level (equalizing them). When the time is up, equalization is automatically terminated.

During equalization, voltage may rise quite high. For a 12-volt system, the voltage may rise as high as 15 to 16 volts. In a 24-volt system, it may rise to 30 to 32 volts. In a 48-volt system, it may rise to 60 volts or higher. Bear in mind that DC loads fed directly from the charge controller that are sensitive to high voltage should be disconnected during equalization. In AC-only systems, the inverter regulates its output voltage to protect loads.

Some experts recommend periodic equalization, but that's not the best advice. When and how often batteries need to be equalized depend on how they are used. If your batteries are frequently deeply discharged, or if composed of multiple series strings in parallel, you may want to equalize them once a month. Or, you may equalize them frequently only during high-use periods. I found that batteries in my off-grid wind and solar system in Colorado worked much harder in the late fall, winter, and spring—the cloudier time of the year. They were more likely to be deeply discharged, so I equalized more often during this period—every month or two. In the late spring, summer, and early fall, I tended to equalize a lot less frequently, unless we experienced unusually cloudy weather during those periods. If you live in a very windy location and your batteries are rarely deeply discharged, they'll need less frequent equalization. For example, batteries that are kept pretty full and rarely or never discharged below 50% may only need to be equalized every six months. Needless to say, there are no hard and fast rules. You have to stay tuned to the weather and the demands on your battery bank, and equalize as necessary.

The frequency of equalization also depends on your normal charge set points. Some homeowners run their battery banks "hot"—at a higher voltage setting than the standard. As a result, the batteries have less need of equalization, but more watering.

Rather than second-guess your battery's need for equalization, it is wise to check the voltage of each battery (using a voltmeter), every month or two. If you notice significant differences in battery voltage in a battery bank—that is, the voltage of one or two batteries is substantially lower than others—it is time to equalize the battery bank. Checking voltage requires a voltmeter (Figure 8.16). They are typically referred to as *multimeters* because they are designed to measure a variety of different electrical

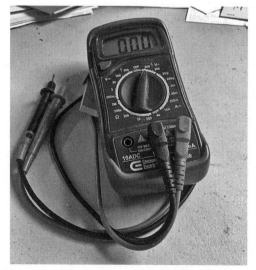

FIGURE 8.16. Voltmeter for Batteries. Multimeters like these can be used to measure the DC voltage of a battery. You can purchase them at hardware stores and home improvement centers as well as electrical supply houses. Don't waste your money on an analog unit. Go digital! Credit: Dan Chiras.

parameters. A digital multimeter is much easier to read than a dial meter. Don't waste your money on the latter.

Another way to test batteries is to measure the specific gravity of the battery acid using a hydrometer (Figure 8.17). Specific gravity is a scientific term for the density of a fluid. Density is related to the concentration of battery acid—the higher the sulfuric acid concentration, the higher the specific gravity. If significant differences in the specific gravity of the battery acid are detected in the cells of a battery bank, it is time to equalize.

Checking the voltage of the batteries and the specific gravity of the cells may involve more work than you'd like and may not be necessary if you pay attention to weather and battery voltage or adhere to a weather-sensitive periodic equalization regime.

Small-wind-system expert Mick Sagrillo recommends using a wind turbine to equalize batteries in off-grid systems, if possible. You'd be amazed at how well it works, as was solar-electric expert Johnny Weiss founder of Solar Energy International in sunny Colorado. Johnny was astonished when he observed a newly installed wind turbine charge a battery bank in a PV/wind hybrid system. The wind turbine, installed during a workshop taught by Mick, was added because the client wanted to wean herself from a gasoline-powered gen-set. After living with the wind/PV hybrid system for a year, the owner sold the gen-set, as she no longer needed it. Bear in mind, this was possible because she had a properly sized PV/wind hybrid system! Unfortunately, conditions have to be right for this to work—that is, you need to have good solar *and* wind resources, and you may need to be willing to curtail usage when both renewable energy resources are scarce.

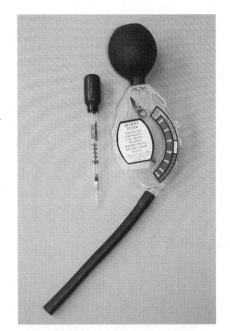

FIGURE 8.17. Hydrometer. Hydrometers like these can be used to test the condition of battery acid, one cell at a time, a process that is time consuming and boring. They measure the specific gravity, which is determined by the concentration of sulfuric acid, as explained in the text. Be careful when you do this so you don't splash sulfuric acid on yourself. Credit: Dan Chiras.

As a final note on the topic of equalization, be sure only to equalize flooded lead-acid batteries. Sealed batteries generally cannot be equalized! If you do, you'll very likely ruin them. (Check with the battery manufacturer, as there are some ways to carefully equalize certain sealed batteries.) Also, don't equalize batteries fitted with Hydrocaps, discussed next. Hydrocaps are installed to reduce battery watering. They must be removed prior to equalization or they will overheat and melt.

Reducing Battery Maintenance

Battery maintenance could take 10 to 30 minutes to an hour each month. (It takes about 10 minutes to check fluid levels in the cells in a dozen batteries but may take a half hour to check them and add distilled water to each cell if battery fluid levels

FIGURE 8.18. Hydrocaps. Hydrocaps shown here are special battery caps that convert hydrogen and oxygen released by batteries into water, which drips back into the cells of flooded lead-acid batteries, reducing watering. Credit: Joe Schwartz.

are low.) If the batteries need a lot of distilled water, your time commitment will be greater. If you have a larger battery bank, say with 24 batteries, you could be in the battery room for an hour or more.

If this sounds like too much work, you'll be delighted to know that there are ways to reduce battery maintenance. One way is to install sealed batteries, although they're best suited for grid-connected systems with battery backup.

Another way to reduce battery maintenance is to replace factory battery caps with Hydrocaps, shown in Figure 8.18. Hydrocaps capture much of the hydrogen and oxygen gas released by batteries when charging under normal operation. Hydrocaps contain a small chamber filled with tiny beads containing a platinum catalyst. Hydrogen and oxygen escaping from the cells react in the presence of this catalyst to form water, which then drips back into the batteries. Because of this, Hydrocaps reduce water losses by 50% to 90% (according to several sources), greatly reducing watering requirements.

Another option is the Water Miser cap. Water Miser caps are molded plastic "flip-top" vent caps. Plastic pellets inside these caps capture most of the moisture and acid mist escaping from the battery fluid, reducing sulfuric acid fumes in the battery room and preventing terminal corrosion. However, they don't capture hydrogen and oxygen. While battery water loss is reduced, and less frequent watering is required, Water Miser caps only reduce water loss about 30%. They can, however, be left in place during equalization.

Of the two, I prefer Hydrocaps. They cost a bit more and must be removed before batteries are equalized (so don't throw out the factory-installed caps; you'll need to put these on during equalization), but they reduce water losses much more than their competitor, which reduces the amount of filling you'll need to do.

Yet another—even better—way to reduce maintenance is to install a battery-filling system, shown in Figure 8.19. Battery-filling systems consist of a series of plastic tubes connected to specially made battery caps (on each cell). Each cap is fitted with float valve (float valves function a bit like the valves in snorkels). In automatic systems, the plastic tubing is connected to a central reservoir containing distilled water. Valves in the battery caps open when the electrolyte level in a cell drops. Distilled water flows by gravity from the central reservoir. When the cell is full, the float valve stops the flow of water. (Remember to keep the distilled water reservoir filled at all times!)

In manually operated systems, the plastic tubing on each string of batteries is connected to a plastic tube fitted with a hand-operated pump. One end is immersed in a container of distilled water, for example, a one-gallon jug of distilled water. A quick coupler allows the tubing connected to the float valves to be connected to the pump and the distilled water supply. Water flows into each cell to the proper level. Pressure buildup tells the operator when all the cells have been filled to the proper level. For a 12-battery battery bank, the entire process takes less than five minutes.

I used a manually operated battery watering system for many years with great success. It is manufactured by Flow-Rite in Michigan. The product is called Pro-Fill and is available online. This battery-filling system converted battery maintenance from a lengthy, boring, onerous task to a pleasure. What used to take me a half hour or more became a five-minute job. Instead of dreading it, and putting off, I eagerly watered my batteries each and every month.

Although battery-filling systems work well, they're costly. A system for 12 batteries in a 24-volt system might cost you as much as a battery. Although a bit pricey, such systems quickly pay for themselves in reduced maintenance time and ease of operation. The convenience of quick battery watering could overcome the procrastination that often leads to damaged batteries, a mistake that can be extremely costly.

A cheaper alternative, though more time-consuming, is a half-gallon battery filler bottle (Figure 8.20). It comes equipped with a spring-loaded spout (similar to the float valves in battery-filling systems). This device allows you to fill each cell—one at a time and very precisely—because the valve shuts off the flow of distilled water when the battery fluid level comes to within an inch of the top of the cell. You can order them from a battery supplier, automobile parts store, or online, for $10 to $20.

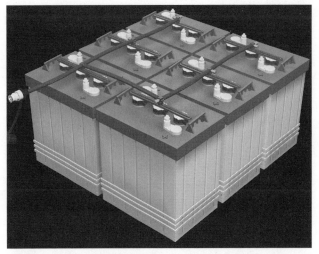

FIGURE 8.19. Battery-filling System. The Pro-Fill battery-filling system greatly reduces the time required to fill batteries and makes the chore much easier. Credit: Jan Watercraft.

FIGURE 8.20. Battery Filler Bottle. Battery filler bottles work well for systems in which batteries are accessible. Credit: Dan Chiras.

FIGURE 8.21. Charge Controller. Charge controllers like the one top left from Apollo Solar regulate the flow of electricity to the batteries in off-grid and grid-connected systems with battery backup. Some charge controllers contain maximum power point tracking circuitry to optimize array output and other features as well, like digital meters that display data on volts, amps, and electricity stored in battery banks. Credit: Apollo Solar.

Charge Controllers

Now that you understand how batteries work and, perhaps even more important, how to take care of them, let's turn our attention to the charge controller, a device that helps us care for our batteries. This information is especially useful for those who install wind and solar hybrid systems. As you shall soon see, in wind systems, diversion charge controllers are the technology of choice.

A charge controller is a key component of most battery-based renewable energy systems. If you're installing an off-grid system or a grid-connected system with battery backup, you'll need one. If you are installing a hybrid solar and wind system, each system will very likely require a different type of charge controller.

A charge controller performs several functions, the most important of which is preventing batteries from overcharging (Figure 8.21).

How Does a Charge Controller Prevent Overcharging?

To prevent batteries from overcharging, a charge controller monitors battery voltage at all times. When the voltage reaches a certain predetermined level, known as the *voltage regulation* (VR) *set point*, the controller either slows down or terminates the flow of electricity into the battery bank (the charging current), depending on the design. In wind systems, charge controllers typically divert surplus electricity to dump loads, described shortly.

Charge controllers have evolved over the years and now come in at least four basic designs. Let's begin with the earliest types of charge controllers: *shunt charge controllers* and *series charge controllers*. Although they are now considered dinosaurs, you may still find them in really old battery-based systems or extremely cheap on the internet. If you are going to add a PV array to your wind system, don't use one of these charge controllers.

As shown in Figure 8.22, a shunt charge controller contains either an on/off switch or a variable resistor. When a charge controller such as this is operating, electricity flows out of the PV array through the charge controller then to the battery bank. When the battery voltage reaches the voltage regulation set point, the shunt is activated. Current flows through the shunt and then back to the array. No more current is delivered to the battery bank. In a switch-type charge controller, the switch closes, completely interrupting the flow of current to the batteries. In a resistor-type charge controller, the resistance to current is high when the charge controller is feeding the battery. As the battery voltage rises, however, the resistance begins to decrease, gradually reducing the flow of current. When the batteries reach the voltage regulation set point, the resistor offers no resistance to electrical current. As a result, all electrical current bypasses (is shunted) the batteries.

Both types of shunt controllers close the circuit or short out PV arrays, which protects the batteries from overcharging. It won't harm the PV modules, however, because they are designed to handle current. (Remember, they are current limited.) How long does the shunt remain operational?

Current flows through the shunt until the battery voltage drops to a predetermined setting, known as the *array reconnect voltage set point*. In a charge controller equipped with an on-off switch, the switch opens entirely, allowing current to flow back to the batteries. In a charge controller equipped with a variable resistor, resistance gradually increases, sending more current to the batteries.

The second type of charge controller is a series charge controller. As shown in Figure 8.23, some series charge controllers contain an on/off switch wired in series. When open, it stops the flow of current to the battery bank.

The series element in a series charge controller can also be a variable resistor that gradually increases resistance as battery voltage climbs toward the voltage regulation set point. This reduces the flow of current to the batteries.

Shunt and series charge controllers were replaced by devices that prevent overcharging in an entirely different manner, known as *pulse width modulation* or *PWM*.

Pulse width modulation (PWM) charge controllers regulate the flow of DC electricity from a PV array to a battery bank by feeding them pulses of electricity of different duration. The lower the battery charge, the longer the pulses. The higher the battery state of charge, the shorter the pulses. In these charge controllers, the length of the pulse determines how much current flows into a battery. As the batteries reach their full charge, the duration of pulses (length of each burst) is reduced. Batteries receive less electricity.

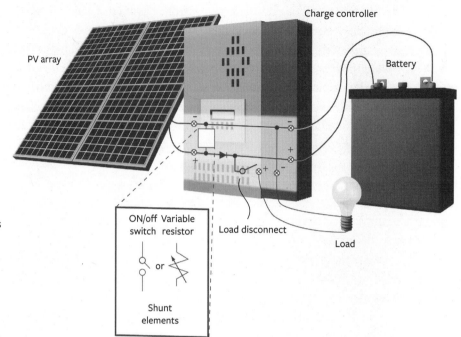

FIGURE 8.22. Shunt Charge Controller. Shunt charge controllers short circuit the array through an on-off switch or a variable resister. Credit: Anil Rao.

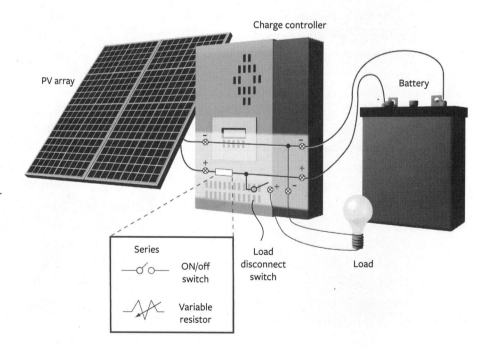

FIGURE 8.23. Series Charge Controller. Series charge controllers disconnect the array, that is, create an open circuit through a series element, either an on-off switch or a variable resistor, as shown here. Credit: Anil Rao.

Pulse width modulation is controlled by a computer that controls a switch between PV array and the battery.

Pulse width modulation charge controllers have themselves been largely replaced by a fourth type, the MPPT charge controller. MPPT stands for *maximum power point tracking*. Like other types of charge controllers, an MPPT charge controller adjusts charging rates based on the voltage of the battery bank, and hence the level of level of charge (how full the batteries are).

Although they perform the same function, these technologies are quite different. By and large, most battery-based systems use MPPT charge controllers. PWM charge controllers are recommended for use in smaller PV systems. MPPT charge controllers are suited for larger arrays. Keep that in mind when installing a PV array to supplement your wind system.

Charge controllers can also be programmed to divert power to auxiliary loads when batteries are full. This feature is used in hybrid battery-based systems that couple wind with PV or micro hydro. They are also used in wind turbine-only systems.

As shown in Figure 8.24, when batteries in these systems reach the voltage regulation set point, the charge controller sends surplus current generated by the wind turbine and/or solar array to a diversion load. A diversion load is an auxiliary load, that is, a load that's not critical to the function of the home or business. In wind-electric systems, it is usually a heating element. Heating elements may be placed inside water heaters or may be in wall-mounted resistive heaters that provide space heat. In windy locations like Wyoming, they can provide quite a lot of additional heat—perhaps all the heat you need to make it through a windy winter.

In hybrid systems, excess power may be available during the summer months as well. In these instances, the diversion load may consist of irrigation pumps or fans that exhaust hot air from barns with livestock.

Why Is Overcharge Protection So Important?

As you now know, charge controllers regulate the flow of electricity to a battery bank in a controlled manner. They also prevent batteries from overcharging.

Overcharge protection is important for flooded lead-acid batteries and sealed batteries. Without a charge controller, the current from a wind turbine or PV array flows into a battery in direct proportion to the output of the turbine or the irradiance, the amount of sunlight striking a solar module. Although there's nothing wrong with that, problems arise when the battery reaches full charge. Without a charge controller, excessive amounts of current could flow into the battery, causing

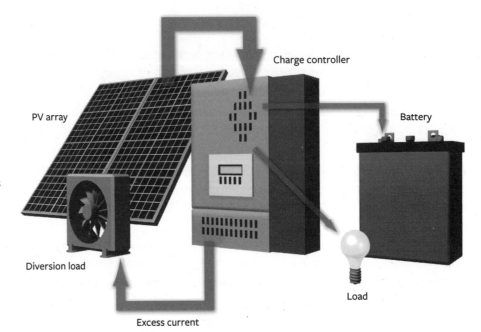

PV array

Charge controller

Battery

Diversion load

Excess current

Load

FIGURE 8.24. Diversion Charge Controller. Charge controllers can be programmed to divert surplus electricity to ancillary (nonessential) loads when batteries are full, allowing a homeowner to get more out of his or her wind or PV system. Credit: Anil Rao.

Preventing Reverse Current Flow

When adding a solar array to a wind system, remember that solar modules are conductors—that is, they not only create electricity, they are capable of conducting electricity. When their voltage is higher than the voltage of a battery bank (when, for example, the Sun is shining) electricity flows from the modules to the battery bank. At night, however, when the PV array is no longer producing electricity, current can flow from the batteries back through the array. That's because the voltage of the battery bank is higher than the voltage of the modules. Voltage pushes electrons. Interestingly, the same theory applies to wind turbines. Current can flow from a battery bank through a turbine if the turbine is inactive and the voltage of the battery therefore exceeds that of the generator.

To prevent this reverse current flow, charge controllers are equipped with a diode in the circuit. Diodes allow electricity to flow in only one direction. This diode is installed in such a way that electricity can flow from the wind turbine or solar array to the battery bank. It cannot flow in reverse.

Were it not prevented, reverse current flow will slowly discharge a battery bank. In most PV systems, battery discharge through the modules is fairly small, and power loss is insignificant. However, reverse current flow is much more significant in larger PV systems. Fortunately, all charge controllers installed in homes deal with this problem automatically.

battery voltage to climb to extremely high levels. High voltage over an extended period causes severe outgassing, which causes batteries to boil, which can lead to the loss of water and sulfuric acid. Water loss can expose the lead plates to air, resulting in their demise. Overcharging can also result in internal heat production. Overheating can cause the lead plates to corrode, decreasing the cell capacity of the battery, which leads to premature death.

Some overcharge is tolerated by a flooded lead-acid battery, so long as the fluid levels don't drop below the top of the lead plates and the electrolyte is replenished. However, overcharging is extremely harmful to sealed batteries. As noted earlier, it can result in a pressure buildup inside a sealed battery that causes water and electrolyte to escape through the pressure-release valve—and there's no way for the water to be replaced.

Overdischarge Protection

Charge controllers protect batteries from high voltage, but also often incorporate *overdischarge protection*—circuitry that prevents the batteries from deep discharging. When the weather's cold, overdischarge protection also protects batteries from freezing. This feature is known as a *low-voltage disconnect.*

Charge controllers prevent overdischarge by disconnecting loads—active circuits in a home or business. Bear in mind that the charge controller only regulates overdischarging of DC loads—appliances, for instance, that run off DC electricity. Figure 8.23 shows the disconnect switch. (As noted in Chapter 7, the inverter prevent overdischarging by AC loads.)

Overdischarge protection is activated when a battery bank reaches a certain preset voltage or state of charge. The low-voltage disconnect not only protects batteries, it protects loads, some of which may not function properly, or may not function at all, at lower-than-normal voltages.

What Else Do Charge Controllers Do?

Although the main purpose of a charge controller in battery-based wind and solar systems is to prevent overcharge and overdischarge of batteries, charge controllers often perform a number of additional functions. In a PV system, for instance, they may control certain loads—that is, switch loads on and off, depending on the time of day. This feature is commonly used to control PV-powered outdoor DC lighting, for example, to illuminate signs or parking lots. In these systems, the solar array essentially becomes a photo sensor. When the current or voltage from the array drops at the end of the day, the charge controller automatically switches on lights.

Some charge controllers may also be wired to outside sensors, for example, temperature or water-level sensors. Temperature sensors allow automatic control of cooling loads (evaporative coolers in greenhouses, for example), and water-level sensors control irrigation pumps.

Finally, charge controllers may control automatic or user-activated equalizations. The charge controller in my off-grid system in Colorado, for instance, has a switch that can be turned on to equalize the batteries from the PV array.

Additional Considerations

When buying a wind system, the manufacturer will provide a controller that works with their turbine and provide instructions on how to install and use it. Shopping for charge controllers for a PV system, however, is a bit more complicated. One thing you will note is that PV charge controllers are rated in amps. Solar charge controllers are rated at either 40, 60, or 80 amps. The amp rating refers to the amount of current a controller can handle from an array. Thus, arrays and controllers must be carefully matched. You wouldn't want to install a 40-amp controller in a PV system with an array that could produce 50 amps under peak Sun conditions. Exceeding the amperage ratings on a controller can overload internal circuits, destroying the unit.

Generators

Another key component of off-grid systems is the generator (Figure 8.25). Generators (also referred to as *gen-sets*) are used to charge batteries during periods of low wind. They are also used to equalize batteries. In addition, generators can be used to provide power when extraordinary loads are used—for example, welding equipment—that exceed the output of the inverter. And, lest we forget, gen-sets may be used to provide backup power if the wind turbine, inverter, or some other vital component breaks down.

Generators are typically "hard wired" to the battery charger inside the inverter. When the generator is started, either manually or automatically, the inverter "waits" about 60 seconds for the generator to warm up and stabilize its AC output. Then it transfers all of the home loads to the generator (a transfer switch is located internally within the inverter). The battery charger then performs a "smart," three-stage charge of the system's battery. It's a good idea, before turning the generator *off*, to terminate battery charging and let the generator run for a few minutes, so it has some idle run time to cool off before being switched off.

Gen-sets for off-grid homes are usually rather small, around 4,000 to 7,000 watts. Generators smaller than this are generally not adequate for battery charging.

What Are Your Options?

Generators can be powered by gasoline, diesel, propane, or natural gas. By far the most common gen-sets used in off-grid systems are gasoline-powered. They're widely available and inexpensive. Gas-powered generators consist of a small gas engine that drives an electric generator. Most generators produce AC electricity. The electricity travels to the inverter containing a battery charger. The battery charger contains a step-down transformer that reduces the 120- or 240-volt output of the generator to a voltage slightly higher than the nominal battery voltage, which could be 12, 24, or 48 volts. The low-voltage AC electricity is then converted to DC electricity. DC electricity is then fed into the batteries.

FIGURE 8.25. Gen-set. Portable gen-sets like these commonly run on gasoline. Credit: Cummins Power Generation.

Gas-powered generators operate at 3,600 rpm and, as a result, tend to wear out pretty quickly. Although the lifespan depends on the amount of use, don't expect more than five years from most heavily used gas-powered gen-set. And you may need to make a costly repair from time to time.

Because they operate at such high rpms, gas-powered gen-sets are rather noisy. If you are interested in quietude, Honda makes some models that are remarkably quiet (they contain excellent mufflers). If you have neighbors, you'll very likely need to build a sound-muting generator shed to reduce noise, even if you install a quiet model. Don't add a muffler to a conventional gas-powered generator. If an engine is not designed for one, adding one could damage it (it creates back pressure on the cylinder).

If you're looking for very quiet and much more efficient generator, consider one with a natural gas or propane-powered engine. Large-sized units—around 10,000 watts or higher—operate at 1,800 rpm and, therefore, are much quieter than their less expensive gas-powered counterparts. Lower speed translates into less vibration and wear and tear. That translates into longer lifespan and less noise.

Natural gas and propane are also cleaner-burning fuels than gasoline. Unlike gas-powered generators, natural gas and propane generators require no fuel handling by you. If yours is fed by natural gas, you never have to worry about running out of fuel, so long as the natural gas service remains intact. If your generator is powered by propane, you will need to keep track of propane levels in your tank.

Natural gas and propane generators are great, but they're more expensive. You could end up paying two and a half to three times more for a natural gas or propane generator than for a well-made gasoline-powered unit. For years, natural gas and propane-powered generators were available only in fairly large sizes. Today, manufacturers are producing a number of lower-wattage generators that could work well in off-grid homes.

Another efficient and reliable option to consider is a diesel generator. Diesel engines tend to be much more rugged than gas-powered engines. As a result, these heavy-duty, long-lasting machines tend to operate without problems and for long periods. Less maintenance means lower overall operating costs and less hassle. Diesel generators are also more efficient than gas-powered generators. Another advantage of diesel engines is that they can be operated on biodiesel, a fuel made from vegetable oil. (It is chemically modified vegetable oil, not straight vegetable oil.) Biodiesel is typically mixed with petroleum diesel fuel in a ratio of 80% petroleum diesel to 20% biodiesel during cold weather. In warmer weather, the mixture can be as high as 99% biodiesel and 1% conventional diesel.

Biodiesel burns cleaner than traditional diesel and is at least partly renewable because it's made from renewable resources—plant-based oils, usually extracted from soybeans or canola.

Although diesel generators offer many advantages over gas-powered generators, they cost more than their gas-powered cousins. And, of course, you will have to fill the tank from time to time. They're also not as clean burning as natural gas or propane gen-sets.

Controlling Noise

William Kemp, author of *The Renewable Energy Handbook for Homeowners*, notes, "Aside from having to run a generator in the first place, the second most annoying feature of a gen-set is having to listen to it running." He goes on to say, "The best way to eliminate this problem is not to operate them at all." Although that's a great idea, it's difficult to practice. In most cases, the best you can do is to minimize run time. Do this by installing a tall tower in the windiest location on your property. Installing a PV array, as noted earlier, to provide additional power and to equalize the batteries, can also help reduce generator run time. But, whatever you do, don't skimp on your generator run time to avoid hassle or eliminate noise. If you do, you could end up sending your battery bank to a premature death.

The next best alternative to not running a gen-set or reducing run time, Kemp goes on to say, "is to locate the unit a reasonable distance from the house and enclose it in a noise-reducing shed."

To dampen noise, insulate the walls and ceilings. However, be sure to create an opening to remove exhaust and provide fresh air. (***Do not* operate a generator inside an attached garage. Dangerous fumes can enter the home.**) I built a shed from leftover 2 × 6s and insulated the walls with a mixture of straw and clay. The ceiling was insulated with a mixture of wood chips and clay. I then applied earthen plaster to the inside and finished the exterior with a coat of earthen plaster with a finish coat of lime-sand plaster. The floor of my shed was made from soil-cement, a mixture of the local subsoil and 5% cement. I installed a louvered fan near the exhaust outlet of the gen-set. The exhaust fan is run off one of the 120-volt AC legs (outlets) on the generator. I created an air intake on the opposite side. This shed was extremely effective in blocking noises and, because of the fan, it stayed cool inside. Never had another complaint from my neighbors!

Kemp recommends building a shed with a floating floor—that is, a floor that does not contact the walls. This keeps engine vibration from radiating outside the building. "The walls should be packed solidly with rock wool, fiberglass or, best of all, cellulose insulation. The insulation should then be covered with plywood or other finishing material, further deadening sound."

Installing a gen-set away from your home may require long-distance overhead or underground feed cable. The cable will need to be sized correctly to reduce line loss. When running the cable underground, it will either need to be rated for burial or installed in conduit. Be sure not to install a gen-set close to a neighbor's house to avoid noise in your own.

A generator with an automatic start switch will make your life a lot easier. This fairly inexpensive feature can be wired to the inverter (for truly automatic function) or to a manual switch inside the home. This feature will save you trips to the generator shed on the freezing cold winter days when the generator is most frequently called into service. Be sure to run wire for the automatic start circuitry when you run the power cable.

As noted earlier, yet another way to reduce noise is to install a quieter gen-set, for example, a gasoline-powered generator with a good muffler, or a lower-rpm diesel, propane, or natural gas generator. If there are no neighbors to bother, it may not be necessary to install a quiet generator or house a gen-set in an insulated shed. However, be sure the generator is protected from weather.

Other Features to Look For

Gasoline-powered generators typically come with four 120-volt AC outlets into which extension cords can be plugged. They also often contain 120-volt 20 or 30 amp outlets that are used to power battery chargers.

Be sure your generator comes with a low-oil shutoff. This prevents the generator from starting if the oil level is low or shuts it off if oil levels fall below a critical level during operation.

Another feature to consider when shopping for a generator is a run-time meter. It will cost a bit more, but will help you keep track of run time so you know when to perform scheduled maintenance.

Living with Batteries and Generators

Batteries work hard for those of us who live off-grid. As you have seen, flooded lead-acid batteries need to be kept in a warm—but not too warm—place. Their enclosures need to be vented, too. These batteries also need to be periodically filled with distilled water. And, you need to monitor their state of charge and either charge them periodically with a backup supply of power when they're being overworked or back off on electrical use. Either way, don't allow batteries to sit in a state of deep discharge for more than a few days if you can help it.

Generators in off-grid systems need a bit of attention, too. If you install one in your system, you will need to periodically change oil and air filters. If you install a manually operated generator, you'll need to fire it up from time to time to raise the charge level on your batteries or to equalize the batteries.

It is also a good idea to run a generator from time to time during long periods of inactivity, for example, over the summer when a generator is typically not used. Gasoline goes bad sitting in a gas tank. It turns into a shellac-like material that can gum up hoses and carburetors. To prevent this, add a fuel stabilizer to the tank at the end of each winter and run the motor from time to time. Finally, gasoline-powered generators can be difficult to start on cold winter days, so be sure to use the proper weight oil during the winter.

Gas-powered generators, while inexpensive, tend to require the most maintenance and have the shortest lifetime. Be prepared to haul your gas-powered gen-set in for an occasional repair. My backup generator visited the repair shop twice in 13 years for costly repairs—and I only ran it 10 to 20 hours a year!

In grid-connected systems with battery backup, you'll have much less to worry about. If you install sealed batteries, for example, you'll never need to check the fluid levels or fill batteries. Automatic controls keep the batteries fully charged. If you install flooded lead-acid batteries, be sure to check them every couple months. Write reminders on your calendar!

Batteries may seem complicated and difficult to get along with, but if you understand the rules of the road, you can live peacefully with these gentle giants and get many years of faithful service. Break the rules, and it's a sure thing you'll pay for your inattention and carelessness.

CHAPTER 9

Maintaining a Wind-Electric System

"Wind turbines are machines. They have moving parts," notes John Hippensteel owner of Lake Michigan Wind and Sun. "That means they require maintenance."

Maintenance takes time and money.

Just as your car, home, and bicycle require occasional maintenance, so does a wind energy system. You wouldn't expect to drive your car for a year without changing the oil, rotating the tires, or performing other basic maintenance and repair. Ignoring these tasks would surely reduce performance and would very likely lead to even more costly maintenance in the future—or even catastrophic failure.

While regular maintenance is vital to cars, it is even more important for wind generators because they work longer and harder in more severe environments.

This chapter discusses maintenance of wind energy systems except for batteries, which are covered in Chapter 8. It is intended to apprise readers of a vital, though sometimes overlooked, part of living with a wind-electric system. This discussion is not meant to discourage you from pursuing wind energy, but rather to inform you of the maintenance requirements of wind energy upfront so you know what you're getting into. I want you to embark on this venture with your eyes wide open. Bear in mind that you can always hire a professional to inspect and maintain your system.

Buyer Beware

Wind turbines have become much more reliable thanks to efforts on the part of manufacturers to reduce the number of moving parts and wear points. Some wind energy experts, however, warn us not to be lured by claims of maintenance-free operation for 20 years made by wind turbine manufacturers or installers. No wind generator is "maintenance free," unless you leave it in its box. Even if the turbine requires little or no maintenance, the tower might.

What's Required

Most wind installers recommend an annual or biannual inspection. Although some wind turbines may be able to operate five years or more without maintenance or inspection, don't wait that long.

Mick Sagrillo recommends twice-a-year inspections of wind systems. The first is in the spring after the weather warms, but before thunderstorm season. Thunderstorms are likely the most violent weather your turbine will encounter, and you want to be sure it's in good shape before Nature's onslaught. The second inspection should be in the fall to be sure your turbine is ready for winter. The last thing you will want to do is climb a tower in a 30 mile-per-hour wind at 30 degrees below zero to take care of something that you could have fixed on a nice, warm fall day.

Inspections should be performed on windless days, of course. This makes climbing the tower (in the case of a freestanding or fixed, guyed tower) or lowering the tower (in the case of a tilt-up tower) easier, safer, and less nerve wracking. Wind turbines on climbable towers should be shut down to prevent injury. *Never* perform inspection or maintenance when the blades are spinning. I want you to be around to enjoy wind energy for a long time. As an added safety measure, after you've climbed the tower, tie the blades to the tower so they can't spin, just in case the winds start to blow while you're working on the turbine. Also, be sure to brake a wind turbine on a tilt-up tower before lowering the tower. This may require you to activate a dynamic brake or engage a mechanical brake.

Inspections and maintenance vary with the wind machine and the towers on which they are mounted. Installation manuals that come with the turbine or tower usually provide detailed instructions. Be sure to read them carefully, and call the manufacturer for any maintenance updates and advice. You may also want to attend the Midwest Renewable Energy Association's tower climbing workshop as well as their maintenance and repair workshop. Recognizing that each machine and tower is different, here I am just offering some general guidelines.

Pre-inspection Monitoring

Because wind turbines require periodic maintenance, it's a good idea to periodically listen to your turbine for odd sounds. When the turbine is first installed, pay attention to the sounds you hear—those should be normal operating sounds. Listen to your turbine in various wind conditions, too. The turbine should operate smoothly and quietly. Pay attention to the vibration in the tower. That's normal.

In subsequent months, listen for unusual sounds—for example, clunking, banging, or grinding. They indicate a bad alternator or bearings. Check for unusual vibrations, too. Any unusual sounds coming from the wind turbine should be noted and are reason for an unscheduled inspection.

Climbable Towers

Before climbing a fixed, guyed tower to check the turbine, inspect the guy cables for proper tension. Tighten, if necessary. Most guy cables are tightened by adjusting turnbuckles, like the one shown in Figure 9.1. Check the connections of guy cables to the anchors. Tighten loose nuts. Check the tower foundation and anchors, including bolts that connect the tower to its base. Be sure that the nuts are tight and are not rusted, or corroded, or cracked. If they are, replace them immediately.

While you are on the ground, also be sure to check the bolts on the cable clamps (Figure 9.2). A socket wrench will make the job of tightening them go more quickly.

FIGURE 9.1. Turnbuckle. This ingenious device is used to tighten and loosen guy cables near their attachment to the anchors. Credit: Dan Chiras.

FIGURE 9.2. Cable Clamps. This photo shows several cable clamps on a guy cables. They secure the guy cable to the turnbuckle, which is attached to the anchor. Credit: Dan Chiras.

When climbing a tower be sure to wear a safety harness, preferably a full-body harness, like the one shown in Figure 9.3. Secure all tools and extra parts so they don't fall and injure people or pets below you. Although I discussed safe climbing in Chapter 6, it's worth repeating some of that advice. First, all climbable towers should be fitted with an anti-fall cable, a fixed climbing cable. An anti-fall cam-type mechanism like a Lad-Saf is attached to the safety harness to catch you should you slip while you are climbing. This sliding "climbing car" follows you as you ascend but locks onto the cable to arrest a fall (Figure 9.4). Once you are atop the tower, belt in with your lanyards and disconnect from the anti-fall cable. Be sure to always secure yourself to the tower. When climbing a tower without a safety cable, "always climb using two lanyards in an alternating pattern so that one of them is clipped onto the tower at all times," advises Jim Green.

Be sure to read through the entire inspection and maintenance protocol of your tower *before* ascending the tower to be certain you have all the tools and parts you need. Make a checklist of tools and inspection and maintenance duties on the front and back of a note card. Carry it with you to the top of the tower. You might want to put a duplicate copy in your pocket in case a breeze comes and sends your list on a cross-country flight.

On your way up, inspect all tower hardware at each tower section joint and at each guy station, where guy cables attach to the tower. Carry a wire brush with you and a spray can of galvanizing compound. Use the brush to remove rust and then cover bare or rusted spots with the galvanizing compound. Replace any hardware that is questionable. If the cross-bracing for the tower is welded, check for cracks or rust at the welds as you ascend. Cracks or rust must be attended to immediately. Cracked welds indicate that something is seriously wrong—do not ignore them! Lower the tower and fix them ASAP.

If your tower is equipped with removable step-bolts for the stub tower (the last section of tower that connects to the turbine), be sure to carry them with you when you climb, using a closeable bag. Or better yet, ask your ground crew to raise them to you on a service line. This line can be a quarter-inch nylon cord run through a pulley at the tower top. The first

FIGURE 9.3. Safety Harness. Purchase a full-body harness for climbing towers. Don't go cheap on this safety measure. The climbing harness will attach to the anti-fall cable and to lanyards once you've stopped your ascent. Credit: Dan Chiras.

climber carries the rope and pulley up and then securely attaches the pulley to the top of the tower. The rope and pulley can be used to raise and lower parts. If your tower is equipped with a work platform, check to be sure it is firmly attached to the tower before stepping onto it.

When climbing a tower or working on a platform, be sure you are securely tied in or clipped in at all times. Experts recommend tying in with at least two lanyards to secure you to the tower once you've reached the top. As John Hippensteel says, one lanyard is for his wife and the other is for his children. Always employ a ground crew of at least one person who is there at all times to service you and other members of the climbing crew, if any. This can save hours of climbing up and down the tower, and is imperative for safety.

When you've reached the top of the tower, check the condition of all bolts that attach the turbine to the stub tower. Tighten, if necessary. Replace, if damaged. Then check the condition of the bolts that hold the blades in place. Tighten loose bolts. Next check the tail vane bolts (the bolts that attach the tale vane to the boom) and any other fasteners. Vibration is a way of life with wind generators, and it causes bolts to loosen over time. You may need to use a torque wrench to tighten bolts properly. Manufacturers include specifications in their installation manuals for proper tightening of all bolts. Be sure to include these torque specifications on your list.

Vibration can also crack metal at bolt holes. Pay particular attention to thin metal parts such as the tail vane. If cracks are present, make a note to order a replacement. Cracked parts on a wind turbine never heal themselves, regardless of how much time you give them.

The blades undergo the greatest wear and tear of any part on a wind turbine, regardless of whether they are made of fiberglass, plastic, wood, or a composite (Figure 9.5). If your wind machine has painted blades, you may need to remove, re-sand, and repaint or refinish the blades to protect them from the elements. (This is obviously a job to be performed on the ground.) If the paint has cracked or the leading edge tape has torn away, the exposed blade material will begin to erode due to abrasion. Moreover, moisture that penetrates the blade can cause the rotor to become unbalanced, severely stressing the wind generator. If your blades show considerable wear

FIGURE 9.4. Anti-fall Cable. Climbable towers should be equipped with an anti-fall cable. Be sure to use it. Credit: Dan Chiras.

Read Before Your Climb

Be sure to read through the *entire* inspection and maintenance protocol *before* ascending the tower to be certain you have all the tools and parts you need. Make a checklist of duties in order. Write big enough so that you can read the list easily. Print the list on one card that fits neatly into your pocket. Carry a spare in case the first one is dropped or swept away by a breeze.

FIGURE 9.5. Check Those Blades. Wind turbine blades often need repair or complete replacement after 10 years or so. Check them very carefully for cracks and signs of wear (especially along the leading edge of the blade) each time you inspect your turbine. Be sure to check the bolts that attach the turbine to the tower and the condition of the wires. Credit: Dan Chiras.

and tear, you may want to perform this maintenance more regularly, for example, every six months.

If your turbine has plastic blades, check for cracks, especially near the base of the blade—the point of connection to the hub. This is a very high-stress area. Check the condition of the leading edge tape, if any. If necessary, re-apply leading edge tape to protect blades from abrasion caused by blowing dust or insects. Check the tips and trailing edge for damage, too. If damaged, repair the blades immediately. After ten years, the blades may need to be replaced, depending on how attentive you were to the first signs of erosion on the blades. Leading edge tape and fiberglass used to repair a slightly damaged blade will look cheap compared to replacing a $6,000 set of blades!

While you are at it, check the blades for tracking by measuring the distance from the stub tower to the tip of each blade. You don't want the blades to strike the tower as they bend under wind loads. Contact the manufacturer for tolerance and shim blades at the root if necessary. (Shimming helps maintain the proper distance between the blades and the tower.)

After checking the blades, examine the alternator of the turbine—if you can get to it. (That's rarely possible.) Look for cracks or damage on all parts of the generator or alternator. Damage here is rare, but it is worth looking for. Cracks in any part of an alternator are an indication that something is seriously wrong. Replace the damaged part before running the wind turbine again.

Next, check the alternator brushes, if any. They should allow the rotor to move easily. (However, most modern turbine alternators are brushless.) Check the alternator bearings to be sure the seal is intact and no grease has leaked out. (This may be impossible to do with sealed turbine bodies.) Replacing a damaged bearing that is starting to rumble is far cheaper than replacing the entire alternator because the magnets and windings grind each other into oblivion.

Once you have inspected the blades and alternator for damage, it is time to inspect other internal components—if you can get to them. Your examination will require a high degree of familiarity with all the components, such as the slip rings. If you are unfamiliar with them, be sure to carefully study diagrams *before* you ascend the tower for inspection. (Here's where an experienced inspector who's familiar with your turbine make and model is worth his or her weight in gold!)

Open the access plate, if your turbine is equipped with one. Begin by checking for obvious signs of corrosion. Check the slip rings for damage or for oil that may have leaked from yaw bearings, the component that allows the wind machine to turn on the tower as the wind changes direction. Inspect all parts for cracks or wear. In a few wind turbines, you may need to grease certain parts or change the oil in the gearbox, if your turbine has one. The Jacobs 31-20, for instance, is a rare turbine that requires oil changes—one each year. Most don't.

After you have inspected the internal parts and closed the access panel or cover plates, check the furling mechanism. If a cable is used to manually furl the tail, inspect the cable. Check its attachment to the tail boom. Look for fraying (unraveling) of the cable. Next, check for cracks and loose nuts in the tail assembly and hinges. Be sure to check the tail pivot pin, pin retainer bolts, and tail pivot bushings (these are all parts that allow "furlable" tails to fold). Examine the surfaces that contact each other at each end of the furling range. These are common wear points, and may need paint, at a minimum.

If overspeed protection is provided by a vertical furling mechanism or a blade pitch mechanism, be sure to inspect the bolts, pins, and springs carefully for damage. Tighten loose bolts as per manufacturers' recommendations and grease as needed.

Finally, be sure to check the condition of exposed wires. Open junction boxes, if any, and be sure all wires are securely attached to one another and that grounding is intact.

With your inspections and maintenance complete, it's time to head down the tower. As you descend, make sure that the electrical wire running from the wind turbine to the junction box at the ground is secured to the tower. Replace fasteners and locking nuts as needed, especially any that have rusted or cracked. If your turbine has a furling cable to shut the machine down for inspections or to protect it from high winds, check the condition of the cable as you descend.

When you have reached the base, be sure the furling winch, if any, is in good shape and operating correctly. Also, check the connections on ground rods (Figure 9.6). Be sure

FIGURE 9.6. Grounding. Excess guy cable is woven through the other cables and attached to a ground wire, attached to a ground rod, not shown here. Be sure to check all of them. Credit: Dan Chiras.

FIGURE 9.7. Clear Before You Inspect. Grasses and other vegetation, including shrubs and trees, often grow up along anchors, making it difficult to inspect guy cables and grounding. You'll need to clear this stuff out of your way first. Credit: Dan Chiras.

FIGURE 9.8. Check the Turbine. Once the turbine is at ground level, be sure to check it and its attachment to the tower very carefully. If your turbine is equipped with remote communications, be sure to check the antenna. Credit: Dan Chiras.

that all contact surfaces are clean and free of oxidation. Inspect other electrical components such as surge arrestors for damage and, if necessary, replace them. If your system is equipped with a dynamic brake, a mechanism that shorts the alternator to stop the turbine, be sure to check the switch.

Tilt-up Towers

If your turbine is mounted on a tilt-up tower, you will need to lower it to inspect the turbine. However, before you begin that procedure, check all of the guy cables where they attach to the anchors. Check for loose or rusted cable clips or hardware and tighten or replace as needed. Check the tilting hinge at the tower base and the attachment of the gin pole to the tower. Make sure that the ground rods are *securely* attached to the guy cables. Bear in mind, you may need to perform some weed whacking or trim shrubs that have grown up around your anchors (Figure 9.7).

After you lower the tilt-up tower, walk along the length of the tower to check the guy cable attachments to the tower. Tighten or replace cable clips or hardware as necessary.

Finally, check the wires coming out of the bottom of the tower to make sure that they are not chafing or chewed up by rodents. Once you are satisfied that the tower is in good shape, inspect the wind turbine as outlined for climbable towers (Figure 9.8).

Final Inspections

When the inspection is complete and you are off the tower (or your tower is tilted back up), you can turn the wind turbine back on. As soon as the machine is operating, listen once again for unusual sounds—for example, clunking, banging, or grinding. They indicate a bad alternator or bearings. Check for vibrations. As noted earlier, unusual sounds emanating from your wind turbine should be noted any time it is operating—and trigger an inspection. Don't procrastinate. The turbine should operate smoothly and quietly.

Last, but not least, you need to check the inverter and/or charge controller to be sure they are working properly. Check for dust or moisture and eliminate the sources.

You also need to test the electrical production of the turbine to ensure that the turbine is working properly. Detailed instructions should be available in your installation manual. Contact the manufacturer if you have any questions on test procedures. Be sure to talk to engineers, not sales personnel. I like to install an additional utility meter on all my wind and solar systems and check it every year at the same time to be sure the turbine is producing what I'm expecting (Figure 9.9). As an added note, I also like to install a 120-volt AC outlet at the base of the tower in case a homeowner or service personnel need to plug in power tools to service or repair the system (Figure 9.10).

Conclusion

Routine inspection and maintenance of a wind machine and tower is a kind of insurance policy. Like taking your car in for oil changes and other scheduled maintenance, it helps ensure longer lifespan and can save you a lot of money over the long haul. Tightening a loose nut is a lot cheaper than replacing a part, or worse, an entire wind turbine that's been damaged when a nut comes off. Procrastinating could create a time bomb, a wind turbine that will decide its maintenance or replacement schedule for you.

Proper maintenance can help the system last for 20 to 30 years—helping you get the most out of your investment. It will also help reduce the amount of mechanical sound a wind turbine produces.

FIGURE 9.9. Utility Meter. I install a meter socket (or base) with each solar and wind system so I can monitor the output of the systems, even if remote monitoring is being used. (It's a good backup should the remote monitoring system fail, as it did with my Skystream after a couple of years.) In this installation of a Skystream 3.7, which produces grid-synchronous AC, the meter is installed at the base of the tower. In systems that require inverters mounted inside or on buildings, the additional utility meter must be mounted after the inverter, and before the main panel or connection to the utility network. That's an AC disconnect next to the utility meter. Credit: Dan Chiras.

FIGURE 9.10. Nice Perk. I install a 120-volt AC outlet (rated for outdoor installation) on all my solar and wind systems at the turbine or array, if possible. Credit: Dan Chiras.

The tower and wind turbine are not all that require maintenance. If you install a battery bank, you'll need to maintain the batteries. Off-grid battery banks typically require a lot more care than batteries installed in grid-connected systems because the latter typically contain no-maintenance sealed batteries. Details of battery maintenance are discussed in Chapter 8.

In contrast, inverters are generally fairly maintenance free. The only attention they require is when they break down, which is rare.

As should be clear by now, routine maintenance and repair are essential to all wind energy systems. If you're not up to the task or can never seem to get around to chores, you should hire a professional to inspect your turbine, tower, and other components every year. If that's not an option, you may want to think about a less maintenance-intensive renewable energy system, like a solar-electric array.

CHAPTER 10

Final Considerations

Zoning, Permits, Covenants, Utility Companies, Insurance, and Buying a System

For those of us who love the idea of generating electricity from the wind, there are few things, if any, in the world more rewarding than raising a tower with a wind turbine and watching the turbine's blades spin for the first time, generating electricity from a clean, abundant, and, in some locations, reliable source of energy. The road from conception to installation of a turbine on a tall tower on your property, however, can be long and arduous. The steps are summarized in sidebar on the following page. As you can see, I have already covered the first three.

In this chapter, I'll discuss most of the remaining steps. I'll begin by discussing building permits and zoning issues, and then turn your attention to interconnection agreements with local utilities, a subject relevant to those who want to install a grid-connected system. Next, I'll discuss concerns neighbors might have and ways to work with them. I will also cover insurance requirements. I'll conclude by providing advice on buying a system and locating a professional installer.

Zoning, Building Permits, and Restrictive Covenants

The decision to install a wind system often involves more than an economic analysis to determine if a wind generator can produce electricity cheaply enough to make the investment worthwhile. There are several additional matters you'll need to check into first. For one, you need to check into zoning regulations—potential legal restrictions imposed by local or state government. These may include height restrictions or setbacks—the height of towers and/or how far a turbine must be placed from property lines and utility lines, respectively.

Steps to Implement a Small Wind Energy Project

1. Measure your electrical consumption.
2. Assess your wind resource.
3. Select a turbine and tower.
4. Check zoning regulations and homeowner association regulations.
5. Check building permit and zoning requirements.
6. Check covenants.
7. Contact the local utility and sign utility interconnection agreement (for grid-connected systems).
8. Obtain building and electrical permits.
9. Order turbine, tower, and balance of system.
10. Install system.
11. Commission—require installer to verify performance of the system.
12. Inspect and maintain the system on an annual basis.

Zoning Regulations

Zoning laws regulate how parcels of land can be used and are common in many parts of the world. Although zoning regulations are in place in numerous cities and towns in North America, some regions, specifically rural areas, have no zoning regulations whatsoever. You're free to do whatever you want. (Often, these regions have the authority to zone, they just don't do it.)

Wind turbines installations are typically subject to zoning laws, so *before* you purchase a system, be sure to check local zoning regulations to be certain it's legal to install a wind turbine, and, if so, what restrictions may be in place.

In cities and towns with zoning, wind turbines may or may not be allowed. If they are allowed, you'll all too frequently find that zoning regulations limit tower height to 35 feet, which is useless when it comes to a wind generator.[1] To obtain permission to install a taller tower, you'll need to obtain a variance or a special use permit—permission to legally vary from zoning regulations. These are issued one property at a time. Developers receive variances all the time, so don't be daunted by this requirement.

Obtaining a variance or special use permit may take several months and can cost thousands of dollars. You'll need to submit a formal request and attend a hearing and may need to hire an attorney to assist in the process. If others have been granted a variance to install towers in your jurisdiction, all the better. If you are installing a system through an established company, they may have already laid the groundwork. That is, they may have already been granted variances for other projects, which could make your request a snap. Or, better yet, they may have worked to change the local zoning regulations, so that you don't have to worry about height restrictions.

If you are in an area in which no turbines have been installed and restrictive zoning regulations are in place, "knowing someone on the city or town council, or even in the mayor's office, can take months, even years, off this process," notes Aaron Godwin, an installer of small wind energy systems.

One problem that applicants for residential wind turbines often come up against is that local zoning regulations do not have *any* stipulations covering small wind systems. As a result, local officials may consider the request under other sections of local zoning regulations. "This is particularly the case when it comes to tower height," notes Mick in one of his many articles on zoning and building permits in *Windletter*. "Wind generator towers are sometimes treated like any other structure on a homeowner's property, and, unfortunately, often severely restricted in height." In many areas, zoning regulations limit building height and, by default tower height, to 35 feet.

Building height restrictions have been on the books forever! And many Code officials have no idea why. I'm told they emerged many years ago to protect factory workers jammed into substandard factories—wooden structures with poor electrical wiring. Because of limitations of early fire-fighting equipment, which could only pump water about 32 feet, early codes set a 35-foot height for buildings. This number was grandfathered into many building codes and still exists in many parts of the United States.

Bear in mind that variances are often granted to permit construction of tall silos, radio and transmission towers, and cell phone towers. "Wind turbine towers," Mick Sagrillo asserts, "should also be exempted from height restrictions, as they become ineffectual at or below the tree line." He adds, "Once zoning committees understand the reasons behind tall towers for residential wind systems, most will entertain height exemptions."

Structural and Electrical Permits

If you are planning on installing a wind turbine in the United States, you may be required to obtain a permit from the local building department. Most projects in areas that call for permits are required to have an electrical permit. In other areas, a structural permit may be necessary for the tower and foundation.

Building permits ensure that the project (1) is safe, (2) is on your property and within required setbacks, and (3) complies with local ordinances, including zoning, if any. They also ensure that your improvement is added to your property records for property tax assessment. Building permits come with a cost. Fees vary from $50 to $6,000, depending on the jurisdiction. How do you find out about potential legal hurdles like permits?

"The quickest way to determine the local codes and requirements is to call or visit the office of the building inspector," note the folks at Bergey WindPower. If you are planning on installing your own system, and it's legal in your area to do so, the more you know about the project, the better. A solid knowledge of electrical wiring is important. If your wind system is being installed by a professional, they'll take care of the permit, and pay the fees.

Building permits are issued after a review of plans that include a site map (drawn to scale) that indicates property lines and dwellings, including nearest neighbors and the location of other buildings. Site maps also show topographic features, easements, if any, and the location and height of the proposed wind machine.

Permit applications also indicate the kind of turbine you'll be installing as well as the tower's height and location. It's always a good idea to submit technical information on the wind turbine you are installing. You may also have to submit data on noise production. Fortunately, thanks to certification by the SWCC, this data is becoming more readily available.

Tower and foundation details may also be required—it all depends on how thorough your building department is. Be prepared to submit drawings of the tower and tower foundation and anchors. Homeowners may also be required to submit an engineering analysis of the foundation and tower to demonstrate that both are structurally sound and comply with local building codes. Engineering analyses are typically provided by a tower's manufacturer and may be stamped by their own engineer. (Their stamp is known as a *dry stamp*.) Their staff are also often very willing to speak directly with local building department officials.

Some municipalities require a *wet stamp*—a stamp of approval by a state-licensed engineer, which could cost $500, possibly more. Check requirements in advance. If this is the case, you may want to appeal the requirement. As Mick Sagrillo points out, "Requirements for a 'wet stamp' cannot be justified. Any company selling towers for wind systems has to perform the engineering analysis to be in business and secure liability insurance." To learn more about tower engineering for building permits, see Mick's article in AWEA's *Windletter* (May, 2006).

Even if not required, it is always useful to submit drawings of the tower and the tower footings when requesting a building permit for a wind system. "A good detailed blueprint will go a long way toward assuring zoning officials that your project is well thought out," according to Mick Sagrillo. Even though the blueprint may not state it, you can assure the building department officials that commercially available towers conform to the requirements of the Uniform Building Code, often referred to as the UBC. If not, they would not be sold by the company.

If the zoning regulations require setbacks—placement of a tower a certain distance from property lines, streets, and overhead utility lines—the site plan and building permit must indicate your compliance with them. Setbacks are required in many cases so that if a tower collapses, it will fall within the bounds of your property. (Tower collapse is extremely rare and greatly blown out of proportion.)

Although setbacks can affect the placement of a wind turbine, don't let them deter you. "Few 600-foot cell phone towers are smack in the middle of a 1,200-square-foot lot, and utility transmission towers are only required to have a narrow right-of-way. If you can find any such exceptions in your area, you have a precedent in your favor," advises Mick. Better yet, he says, talk with your neighbors about setbacks. "Most ordinances are written so that if your neighbor has no objection to siting the tower closer to the property line, the permit can be granted."

Electrical permits are commonly required to ensure the system is installed according to local electrical code to ensure safety. Local officials use the National Electric Code (NEC) to govern the installation of all wiring and associated hardware such as inverters. "The NEC is in place to protect against shocks and fires caused by electricity," notes Mick. "If you ever have a problem that involves an insurance claim and the system was not Code compliant when it was installed, it is possible that your insurance company could refuse to honor your claim." Be sure to inquire which version of the Code they use, for example, 2014 or 2017. In rural areas, you may find that older versions of the Code are used. In more up-to-date areas, the newest version is always applied.

Even if the building department does not require an electrical permit and inspections, the utility may want to inspect the system if you are connecting to the grid. "It, too, will be looking at whether the system is NEC compliant," notes Mick. "Quite often the utility will require that you submit a single-line diagram of the system designating all electrical components." Utilities will also require you to submit a pre-construction agreement, signed by an electrician or engineer who has reviewed the plans. It stipulates that the system will be installed to Code. Utilities also require a post-construction agreement signed by an electrician or licensed engineer saying that the system was indeed installed safely and according to Code.

In rare instances, Mick Sagrillo has seen local building departments require homeowners to employ a licensed wind system contractor or installer. "No such system of licensing or certifying of wind installers is in place anywhere in the US at this time, so no one can meet this requirement." You should appeal such a requirement on these grounds.

Building department inspectors will visit your site at various stages to ensure that you are doing things correctly. An electrical inspector, for instance, will most likely visit two times. The first visit is to inspect underground wire runs that often go in early. Be sure to leave the ditch open so they can verify the depth of your wire. Their second visit occurs after the turbine and tower are in place, to inspect final wiring. A structural inspection may also occur at that time. Be sure to ask what inspections your building department requires and what they'll be looking for.

One thing you will find in rural areas is that one inspector may be assigned structural and electrical inspections. Their next job may be a plumbing inspection on a home under construction three miles down the road. My experience is that these inspectors are pretty knowledgeable, but there are gaps in their knowledge, especially as it pertains to renewable energy systems. Respect these individuals, but politely challenge them if they make an unfair assessment. It's best to be on site when an inspector comes to visit so you can answer questions.

When applying for a building permit, be sure to make it clear that you are planning on installing a small, residential wind machine to offset your own personal electrical demand. Be very clear on this point from the get-go. Don't just announce that you want to install a wind turbine on your property. The building department may think you are planning to install a huge commercial wind machine.

"In spite of all potential hurdles, never approach such interactions with a chip on your shoulder," Mick Sagrillo advises. "You can challenge an unfavorable decision or ruling in court, but you will only make enemies in the process, and spend more money that you thought possible." However, do not let anyone deny you the permits you request based on uninformed contentions. Always remember that when it comes to permitting, the burden of proof is on the building department, not you.

"Never go to meetings with a pugnacious attitude or a list of demands," advises Mick. "From the official's perspective, there is no bigger turnoff than having to deal with an arrogant know-it-all." On the other hand, don't be intimidated by resistance when dealing with building code officials or at public hearings, if any, provided you know what you are talking about. You should expect to be treated in a reasonable and timely manner by everyone involved. Quite often, just letting a building department know that you have consulted with an attorney about your rights as well as your responsibilities (which may be a good idea anyway) will go a long way in reassuring them that you know what you are doing.

Finally, be aware that building departments and electric utilities have tried to dissuade people from installing wind generators by employing stalling tactics, notes Mick: "To avoid delays, set mutually agreed-upon deadlines for decisions, with homework assignments for each party as appropriate."

And while we are on the subject, avoid the tendency to bypass the law—that is, not obtaining a building or an electrical permit. This can create huge problems. Municipalities have the legal right to tell a homeowner to remove his or her wind turbine and tower if the homeowner does not secure a required building permit. In fact, there are some noteworthy examples of municipal governments that have forced homeowners to remove expensive unpermitted wind turbines and towers. My advice: always obtain all required permits. Do not, under any circumstances, bypass the requirements. The consequences are just too great.

The same applies when connecting to the grid. Be sure to follow the utility's procedures to the letter. Mick Sagrillo knows people who have installed grid-connected systems without permission and have been denied access to the grid by the local utility company. As a result, they were not able to run their systems.

Obtaining a permit for a wind system may take several months, so apply many months in advance, *and don't buy a wind energy system until your permit has been granted.*

Covenants and Neighbors' Concerns

Covenants from homeowner associations can also create a huge obstacle to wind system installations. Even with permission from the zoning department and a permit from the local building department, restrictive covenants in force in many neighborhoods can block an installation. Covenants govern many aspects of peoples' lives—from the color of paint they can use on their homes to the installation of privacy fences. They often expressly prohibit renewable energy systems, such as solar hot water systems and solar-electric systems, although they may be silent on wind systems, because these systems are rarely installed in neighborhoods.

Unfortunately, courts have consistently upheld the legality of covenants imposed by neighborhood associations. If you live in a covenanted community, contact the head of the homeowners or neighborhood association. Covenants are typically overseen by a architectural review committee—a volunteer group that reviews applications for changes to homes, including installation of solar and wind systems. The architectural review committee will explain the process required to apply for permission to install a wind turbine and tower.

Even in the absence of restrictive covenants, you should consider the needs and desires of your neighbors. "Many people feel strongly about the need to preserve the landscape, views, history, and peace and quiet of their neighborhoods," writes Steve Clarke, Canadian wind energy expert. So, unless you live on a large piece of property, be sure to discuss your plans to install a wind turbine with your closest neighbors long before you lay your money down.

Don't expect your neighbors to be as enthusiastic about a wind turbine on your land as you are. Not everyone is enamored of wind energy. Some people seem to have a knee-jerk reaction against wind turbines.

Many of them simply fear the unknown. Others seem to object to anything new or environmental. Some neighbors may object out of spite or because of unresolved anger. "If you are feuding with your neighbors, make up with them," advises wind expert Robert Preus. Even though your neighbors may have no legal recourse, they can make your life miserable. They could turn up at zoning hearings and voice their opposition, blowing concerns out of proportion.

Although many concerns about wind systems are misguided, be prepared to respond to all concerns without being defensive. You may want to show your neighbors pictures of the wind turbine and tower to allay their fears. If you are good at Photoshop, you can even take a photo of your home from their house and place a picture of the wind turbine and tower on your property so they can see what it will look like. If you are thinking about installing a grid-connected system, let them know that you'll be supplying part of their energy, too—when your system is generating a surplus. People seem to get a kick out of that.

One concern that's terribly misinformed is bird kills. As noted in Chapter 1, most of the concern about bird mortality is from commercial wind farms and, even then, it is grossly exaggerated. You may want to cite the statistics presented in the first chapter of this book about the perils birds face from cats, windowpanes, automobiles, telecommunications tower, and pesticides and how little risk wind machines pose to our fine feathered friends.

Some neighbors may be concerned about interference with radio and satellite television reception. As noted in Chapter 1, the plastic or fiberglass blades of modern residential turbines are transparent to electromagnetic waves such as radio and television. They do not interfere with reception.

Another concern is sound. Although concerns about sound are important, they too are usually blown out of proportion. At a distance of 100 feet, a well-designed, low-rpm residential wind turbine produces a sound level of about 45 decibels. This sound is lower than the background sound level produced in many homes and offices. Most small wind turbines make less sound than a residential air conditioner when operating (not furling). Share this information with your neighbors.

Although turbine sound emissions increase with increasing wind speed, so does ambient sound. Background noise on a windy or stormy day such as the rustling of leaves and grasses usually drowns out much of the turbine's sound. There's so much going on that few people can hear a wind machine at all. Because most people are inside in such weather, there's even less chance that a neighbor will hear your wind turbine. You may want to take a neighbor to visit a similar wind turbine to illustrate the point.

While you are at it, be sure to check out local sound ordinances, if any, to ensure you'll be in compliance. Let your neighbors know that you are concerned about protecting them from unwanted sound and are doing everything possible to safeguard them—for example, you are mounting your machine on a very tall tower in a location on your property that minimizes sound beyond your property line. You may want to select a wind turbine with a lower rpm. As a rule, the faster the rpm, the louder the turbine.

As a final note on the subject, if you are going to install a freestanding tower, be sure to warn neighbors that you'll be using a crane to install it. That way, they won't freak out when they see a monster crane show up for the installation.

To help you allay your neighbors' fears, some wind turbine manufacturers offer Q & A sheets for residential wind turbine. They address many of the common concerns and can be a great resource when warming your neighbors up to the idea of a wind turbine on your property. You might make a copy of such material to pass out in person to your neighbors. A typewritten summary of key points may work well. Don't leave material in mailboxes. It's very likely not going to be read, unless you hand it to your neighbors personally. Also, resist the temptation to overwhelm them with written material. It's best to talk to your neighbors in person so you can address concerns directly—before they are blown out of proportion. Face-to-face meetings allows you to gauge their reaction and determine if something is bothering them. You may also want to type up a few key talking points and practice them in advance.

Giving neighbors advance notice, answering any questions they have, and being responsive to their concerns is the best way to avoid misunderstandings and problems later, note the folks at Bergey WindPower. "This is doubly good advice if your property size is less than 10 acres or you have to obtain a variance for a building permit. Good neighbor relations boil down to treating your neighbors the same way you would like to be treated and showing respect for their views," they add. "An example of what not to do is to put the turbine on your property line so that it is closer to a neighbor's house than to your own and not give those neighbors any advance notice of your intentions."

Aesthetics and Zoning Hearings

One issue that may come up at zoning hearings for residential wind turbines is aesthetics, a topic that is remarkably difficult to deal with, as Mick notes in an article in *Windletter* (May 2004), adapted here with Mick's permission.

Aesthetics is, of course, highly subjective. Nonetheless, there are steps you can take to make your wind system as visually unobtrusive as possible. As a wind system applicant, you can offer some concessions at hearings to assure neighbors that your wind system will not be an eyesore. Mick's list includes the following:

1. The wind system will not be painted a garish color, like hunter orange or electric chartreuse. Wind turbines are painted by the manufacturer, and their colors have been thoroughly considered from two angles: to make sure that they blend in with the environment and to make them distinctive from other manufacturer's turbines. In practice, the first takes precedence over the second. Manufacturers shy away from painting their products in annoying colors. Some manufacturers may be willing to paint your turbine a different color, so be sure to check with them.

   Towers are most often made of galvanized steel. They come from the factory bright and shiny, but soon weather to a muted gray color, which readily blends in with the sky. Several locations across the country require towers to

Protecting Your Right to Install a Wind System

Securing permission to install a wind turbine on a tall tower can be a time-consuming, and sometimes costly ordeal, if you have obstructionist neighbors and a jittery zoning committee. This isn't always the case, however. In Wisconsin and Nevada, for instance, state law limit a zoning department's ability to place unnecessary requirements and burdens on individuals seeking a building permit for a wind system. Be sure to check with a local installer or your local state representative's office to see if your state has such a law.

be painted green to blend in with the surrounding vegetation. In almost all the circumstances where Mick has seen this required, the green tower stands out far more than does a weathered-gray, unpainted galvanized tower.

2. The wind system will not display any advertising other than the manufacturer's logo, which is usually on the tail or body of the turbine.

3. The tower will not support any signs, other than perhaps some cautionary signs at the base.

4. The turbine will not be flooded with light at night, as is the case with billboards.

Connecting to the Grid: Working with Your Local Utility

If you are installing a grid-connected system, as noted above, you'll also need to contact your local utility company to work out an arrangement to connect to their system. You'll need to assure them that the electricity you will be providing will be of the same quality as grid power—240 volt 60 Hz in North America. Utilities may have questions regarding disconnection, notably what happens if the grid goes down and they need to send a worker to repair it. Is the proper equipment in place to ensure that the wind system will not backfeed onto a dead grid? They don't want their employees getting shocked by a wind system that's operating while their system is down. If you are installing a UL-1741 listed inverter, you're covered. That's all they need to know. The interconnection agreement will also inform you how you will be compensated for any surplus electricity (net excess generation) you will be supplying to the grid. That's determined by state law.

The interconnection agreement may require or suggest that you carry a certain amount of liability insurance—sometimes as high as $1 million. Be sure you read this language very carefully to determine whether liability insurance is mandatory or recommended. (More on insurance shortly.)

According to Mike Bergey, "After over 200 million hours of interconnected operation, we now know that small utility-interconnected wind turbines are safe, do not interfere with either utility or customer equipment, and do not need any special safety equipment to operate successfully." Moreover, he says, "the output of a wind turbine is made compatible with utility power using either a line-commutated inverter or an induction generator." If you need help with such matters, call your installer or the wind turbine manufacturer. They should be able to help allay the fears and address the concerns of your utility.

As for payment, all utilities in the United States are required by federal law (the Public Utility Regulatory Policy Act of 1978) to buy surplus electricity produced by their grid-connected customers. There's no getting around that. How much they pay is a different matter.

Federal law requires utilities to pay the "avoided cost"—that is, what it costs the utility to generate the power, which is usually one-fourth to one-third the amount they charge their customers. If you are paying eight cents a kilowatt-hour, the avoided cost may be as low as 2.5 cents. That's all the utility is required to pay.

But that's not the end of the story. Many states have enacted net metering policies that require utilities to pay residential clients the same price they charge. That is, if a utility is charging ten cents a kilowatt-hour, they pay the same rate for power delivered to them.

In days past, some utilities installed two meters on wind and solar systems, one to track the electricity they sell you, and another to track the electricity you sell them. Most utilities, even rural electric co-ops, are switching over to one-meter systems that digitally track what you put on the grid and what you take off. Check with your local utility to determine their practice and how much you'll be charged *before* you purchase a system. My experience with rural electric co-ops in our area is that they're extremely fair about charging to upgrade a meter. Others may not be so easy to work with.

Contacting Your LOcal Utility

At one time, making initial contact with a local utility could be a rather formidable task, however, these days, with so many renewable energy systems in place, it's pretty easy to find someone at a local utility who can help you. Their interconnection agreement is typically posted online or can be mailed to you. By all means, first check out the utility's website and look for terms such as "net metering" or "interconnection agreement." Be sure, if you perform a search, that you have the latest application. You don't want to spend time filling out an application that is two years out of date.

I've found that some utilities provide lengthy documents that contain interconnection requirements for different types of systems—for example, small residential systems and large commercial systems—that operate under different rules. The info packet can be rather confusing, so give the utility a call if you are perplexed and need a little guidance. Same goes for filling out the application, which is usually just a couple pages long. There may be blanks on the application that don't apply to your installation. Call before you go nuts over such matters.

Under federal law, utilities are required to allow customers with wind generators and other renewable energy systems to connect their systems to the utility grid. "While utilities are well aware that they are so required," says Mick, "getting them to cooperate may be another matter entirely." Problems may occur in dealing with rural electric associations, some of which are hostile to the idea of small-scale renewable energy. As a result, it often pays to do a little advance research. You can learn about your state's net metering policy by logging on to the Database on Incentives for Renewables and Efficiency, www.dsireusa.org.

Working Through the Details

Interconnection agreements are pretty simple. They ask questions about the system, like the size of the turbine, and means of disconnection—both the utility disconnect and the automatic disconnect. As noted in Chapter 3, utilities typically require you to install a visible, lockable disconnect—a manually operated switch that allows utility workers to disconnect your wind system from outside your home in case the grid goes down and they need to work on the electric lines. As you learned in Chapter 7, although grid-connected inverters automatically terminate the flow of electricity to the grid when they detect a drop in line voltage or a change in frequency, the disconnect may still be required by your local utility.

Even though utility disconnects are rarely, if ever, used, do not fight your utility on this requirement, says Mick. "In this and other situations, always remember: utilities often consider 'allowing' you to hook up your wind system to their grid analogous to 'doing you a favor,' even though they are required to by federal law."

Interconnection applications also require a one-line drawing. One of the application's most important purposes is to provide you an opportunity to indicate the anti-islanding feature of the inverter you install. (That's the system that automatically disconnects the inverter from the grid should there be a power outage). I've found that rural electric co-ops usually ask to inspect the system and will test this feature. They will kill the power to the property and measure the voltage from the wind system to be sure it truly has shut down.

When dealing with a utility that is inexperienced with grid connections, do not allow them to charge you thousands of dollars in engineering fees to review your electrical schematics or test your system. Utility interconnection of wind systems has been occurring for 40 years and tens of thousands of such systems are interconnected to the grid across the United States, to say nothing of worldwide installations. This is not a new or unproven technology.

Fostering a Productive Relationship

When working with utilities, be polite, respectful, and cordial. Keep in mind that the utility is a lot bigger than you are, with much more legal expertise and more resources to fight your installation than you could ever dream imaginable. Your utility can make your approval process very fast or unbelievably difficult. It has all the marbles, and it has the bag, too. And it knows it. Few homeowners have the resources to challenge a utility in court. In a case of irreconcilable differences, your best bet is to contact the public utility commission for help.

If you go into the process armed with the proper knowledge, however, it is unlikely ever to come to that, says Mick. It is much better to invite your utility out to inspect your wind system, test the automatic shut-down features, and become a believer. Their staff members will witness for themselves your wind turbine generating electricity and feeding it onto the grid. I often invite the utility to send employees to visit my solar and wind installations—and they usually take me up on the offer. I also offer to train their employees while they are visiting, helping them become more familiar with renewable energy systems. I've found that utility workers are almost always curious about how these systems work—and whether they make sense. They've typically been brainwashed into thinking that renewable energy isn't worth the investment, so it's helpful to have some numbers on the cost of electricity and return on investment ready.

Insurance Requirements

Two types of insurance are required, and they should be in place *before* you install your wind system: homeowner's insurance to protect against *damage to the turbine* and liability insurance to protect you against *damage caused by the turbine*.

Homeowner's Insurance

The most cost-effective way to insure a wind system against damage is under an existing homeowner's policy. This is far cheaper than trying to secure a separate policy for a wind system.

When you contact your insurance company, explain that you wish to insure a small residential wind turbine and tower that will be generating electricity for your home or business. If you really want to impress the agent, tell her or him that you want to insure an "appurtenant structure" on your current homeowner's policy. This is a term used by the insurance industry to refer to any uninhabited structure on your property not physically attached to your home. Examples include unattached garages, silos, barns, storage sheds, grain elevators, satellite dishes, and towers.

Insurance premiums on a homeowner's policy fall into two categories, each with differing rates. Your home is assessed at a higher value than an unattached garage or a storage shed or your tower. This is because people's homes are more lavish than most garages and sheds, and contain myriad personal possessions, furniture, and clothing not typically found in other structures.

Appurtenant structures, on the other hand, are assessed and charged at a lower rate. It is usual practice for insurance companies to insure appurtenant structures for the total cost of materials plus the labor to build the structure. This represents the installed cost of a wind system.

Coverage of appurtenant structures on a homeowner's policy only applies if the system is installed on the same property as your house. If it is on a separate piece of property, you will very likely have to insure it under a separate policy.

It is best to insure a wind system for its full replacement cost, and not a depreciated value over time. Towers do not wear out, and a properly maintained wind turbine can easily last two decades or more. Any wind system should have insurance coverage that includes damage to the system from "acts of God," plus possible options for fire, theft, vandalism, or flooding.

While most wind generator towers are designed to withstand over 120 mile-per-hour winds, tornadoes or hurricanes can wreak havoc on them, just as they would any other structure in their path. Damage may be caused by the extreme wind blowing on the tower and turbine or by debris blown into the tower. Few structures can withstand having a sheet of plywood torn from a garage roof blown into them at 80 miles per hour.

Another "act of God" of concern is lightning strikes. A properly installed wind generator tower has a ground rod connected to each tower leg or at each anchor where a guy cable is connected to the earth, as described in Chapter 6. The wires from the wind generator to the controller and inverter should also be protected by a good lightning arrestor. Grid-connected systems are also grounded on the utility side of the inverter. Adding a lightning arrestor and voltage surge arrestors on the utility side of the inverter affords additional protection. While none of this will guarantee that your system will not be damaged by a lightning strike, it certainly reduces that likelihood. Plus, in the eyes of the insurance company, you have taken prudent measures to protect your system.

Fire is of minimal concern to wind turbines and towers. However, it should go without saying that the wiring of the entire system must comply with the state or local electric code. If you have a fire in the house caused by some funky wiring in the wind system, your claim may be denied and your policy canceled.

Theft of an entire wind system, or even any part of the system, seems unlikely, so should not be a major issue. Vandalism, on the other hand, may be a bigger concern for wind turbine owners. While incidents of vandalism are infrequent, they have occurred. The most frequently filed claim attributed to vandalism involves guns being fired at the turbine's blades or the generator itself. In either case, damage can be substantial.

Flood insurance is a nationally administered program that is usually geared to damage to primary dwellings. Costs can be exceptionally high for a home located along a coastline or in a floodplain near a stream or river with a penchant for flooding. Since this is a very site-specific assessment, no insurance company Mick has contacted would even quote a range of prices. If you live in a floodplain, get an estimate before beginning construction.

Insuring a wind system as an appurtenant structure on a homeowner's policy is relatively inexpensive. While a percentage of the home insurance coverage extends to appurtenant structures, added insurance can be purchased. Since most rural homeowner insurance claims are for fire damage, the deciding factor in pricing coverage is determined by the homeowner's distance from the nearest fire department, regardless of whether your wind turbine can catch fire or not. As a result, the additional premium would likely be slightly lower for a system on property on the outskirts of a town, and slightly higher for a system sited farther away.

Liability Coverage

Homeowner's insurance protects against damage *to* the turbine. Liability insurance protects you, the owner, against damage caused *by* the wind system, for example, if the tower falls on a neighbor's garage. It also protects you from personal injury claims or claims stemming from the death of employees working on a utility line during a power outage. Even though this can't occur because of the protective mechanisms built into inverters, utilities sometimes insist on liability coverage.

Liability coverage is relatively inexpensive; in fact, it comes with a homeowner's policy. In most locales, liability coverage for your home is $300,000. This is the minimum coverage required for anyone with a federally insured mortgage. Increasing this to $500,000 coverage may add an additional $10 to $20 an annual premium in most areas. Extending coverage to $1,000,000 may add $35 to $40 more to the annual premium. It is advisable to ask about an umbrella policy for liability coverage of $1,000,000 or more, as the rates are cheaper still.

If your wind energy system is grid-connected, the local utility may dictate the level of liability insurance that it requires as a condition for interconnection. If the amount seems unreasonable to you, file an appeal with your state's public utility commission. If you live in a rural area, ask the utility what liability insurance requirements they have for farmers with emergency standby generators. Since these systems have similar protective devices, the requirements should be similar. By doing this, you can assure yourself that the utility is not merely upping the ante for your wind system to dissuade you from competing with them by producing your own electricity.

Buying a Wind Energy System

If wind energy seems like a good financial, social, or recreational pursuit for you, I strongly recommend hiring a competent, experienced professional wind site assessor and a professional installer—although they are few and far between. A local supplier/installer with experience and knowledge can be a great ally. They will supply all of the equipment, be certain that it is compatible, help obtain permits, if necessary, and install the system. They'll test the system to be sure it is operating satisfactorily and will be there to answer questions and to address any problems you have with the equipment.

I recommend that you find an installer who also provides routine maintenance—for example, someone who will climb or lower your tower and check things over at least once a year. Be sure to find an installer who stands by his or her work. Most wind turbines are guaranteed for five years, although a couple companies now sell turbines with 10-year warranties. You want to be sure the installer will respond promptly and knows what he or she is doing. Look for people who've been in the business for a while—the longer, the better—and who have installed a lot of systems. Local installers who know the business are worth their weight in gold. Consider buying turbines that have been on the market for a while, too. I'm a lot more confident about a wind turbine that's been in production for 20 years than one that's only been on the market for two years.

As with any major home project, ask for references, and call them. Visit wind installations, if possible, and contact the local office of the Better Business Bureau to find out the installer's rating. Get everything in writing from an installer, too. Be sure to sign a contract. Be sure the installer has insurance to protect employees during installation. Don't pay for the entire installation up front. If possible, be on the site when the work is done. (Just don't get in the way!)

You can also purchase equipment from a local or online supplier, and put it up yourself with or without their guidance. I don't recommend this route unless you are handy and have attended a couple of installation workshops. Raising a tower is risky business. Wiring is fraught with difficulties. Connecting to the electrical grid is a job for professionals. A wind energy system is such a huge investment that you don't want to mess it up.

Another option is to buy from a local supplier or an online source and sponsor a workshop on your property to take care of the installation. Nonprofit organizations like Solar Energy International and the Midwest Renewable Energy Association may be looking for wind energy installations in different parts of the country. They fly in wind energy experts who teach a one- or two-week installation workshop. Workshop registrants pay a fee to cover the cost of the instructor, the costs of advertising and setting up the workshop, food, drinks, snacks, and perhaps even a portable toilet. Their fees do not underwrite materials costs. The homeowner pays the costs of materials. A local contractor may be needed to overseer the preparation and installation. (He'll be the installer of record, required for the permit.)

Although you may not save any money on the installation, workshop installations can be very satisfying. You do have to be comfortable with a dozen or more people on your site for a week. If the workshop leader is competent and checks all of the attendees' work, you'll get a wind system installed while helping a group of people learn about wind electricity. Be sure to contact these nonprofit organizations well in advance and be prepared to help organize and publicize the workshop. Don't call up a month or two prior to the installation date to see if an organization can use your site for training. Most organizations schedule a year out, so summer installations were planned the previous fall. Workshop schedules are usually placed online by the first of the year. Also, be sure your insurance will cover volunteer workers on your site.

Parting Thoughts

You have now reached the end of this book. When you started reading, you may have simply wanted to determine if a wind system was suitable for your home or business. Perhaps you were sure you wanted to install a wind system, but didn't know how to proceed. I hope that you now have a clear understanding of what is involved.

So, what will you do with this new knowledge?

If you have come to realize that your dreams for a wind system were unrealistic, you can pursue other dreams—and be glad that you did not spend a lot of money on a wind system that would not have met your expectations. If, however, you now know that a wind system is right for you, perhaps the next step is to investigate available systems, find a dealer/installer, and start harnessing the power of the wind.

With ongoing inflation and increasing demand for the materials that go into a wind system—steel, copper, cement, rare earth magnets, etc.—prices of wind turbines and towers are not likely to come down. The longer you wait, the higher the cost of installing a wind system is going to be. The cost of electricity and the fuels needed to produce the materials may also be rising. The sooner you get your wind system up and running, the sooner you will start saving.

My job is over. I've helped you get up to speed on wind energy systems. You now know a great deal about wind, wind energy systems, electricity, wind turbines, towers, inverters, batteries, and the maintenance requirements of wind energy systems. You've obtained a good solid working knowledge—good theoretical as well as practical information—that puts you in a good position to move forward. I've given you a lot of advice on installation and helped you grapple with economic issues. My work has ended; yours is just beginning. I wish you the best of luck! May the wind be as much an ally to you as it has been to me.

Net Metering Policies by State

State	Subscriber limit (% of peak)	Power limit Res/Com(kW)	Monthly rollover	Annual compensation
Alabama	no limit	100	yes, can be indefinitely	varies
Alaska	1.5	25	yes, indefinitely	retail rate
Arizona	no limit	125% of load	yes, avoided cost at end of billing year	avoided cost
Arkansas	no limit	25/300	yes, until end of billing year	retail rate
California	5	1,000	yes, can be indefinitely	varies
Colorado	no limit	120% of load or 10/25*	yes, indefinitely	varies*
Connecticut	no limit	2,000	yes, avoided cost at end of billing year	retail rate
Delaware	5	25/500 or 2,000*	yes, indefinitely	retail rate
District of Columbia	no limit	1,000	yes, indefinitely	retail rate
Florida	no limit	2,000	yes, avoided cost at end of billing year	retail rate
Georgia	0.2	10/100	no	determined rate
Hawaii	none	50 or 100*	yes, until end of billing year	none
Idaho	0.1	25 or 25/100*	no	retail rate or avoided cost*
Illinois	1	40	yes, until end of billing year	retail rate
Indiana	1	1000	yes, indefinitely	retail rate
Iowa	no limit	500	yes, indefinitely	retail rate

State	Subscriber limit (% of peak)	Power limit Res/Com (kW)	Monthly rollover	Annual compensation
Kansas	1	25/200	yes, until end of billing year	retail rate
Kentucky	1	30	yes, indefinitely	retail rate
Louisiana	no limit	25/300	yes, indefinitely	avoided cost
Maine	no limit	100 or 660*	yes, until end of billing year	retail rate
Maryland	1500 MW	2,000	yes, until end of billing year	retail rate
Massachusetts**	6 peak demand 4 private 5 public	60, 1,000 or 2,000	varies	varies
Michigan	0.75	150	yes, indefinitely	partial retail rate
Minnesota	no limit	40	no	retail rate
Mississippi	N/A	N/A	N/A	N/A
Missouri	5	100	yes, until end of billing year	avoided cost
Montana	no limit	50	yes, until end of billing year	retail rate
Nebraska	1	25	yes, until end of billing year	avoided cost
Nevada	3	1,000	yes, indefinitely	retail rate
New Hampshire	1	100	yes, indefinitely	avoided cost
New Jersey	no limit	previous years consumption	yes, avoided cost at end of billing year	retail rate
New Mexico	no limit	80,000	if under US$50	avoided cost
New York	1 or 0.3 (wind)	10 to 2,000 or peak load	varies	avoided cost or retail rate
North Carolina	no limit	1000	yes, until summer billing season	retail rate
North Dakota	no limit	100	no	avoided cost
Ohio	no limit	no explicit limit	yes, until end of billing year	generation rate
Oklahoma	no limit	100 or 25,000/year	no	avoided cost, but utility not required to purchase
Oregon	0.5 or no limit*	10/25 or 25/2,000*	yes, until end of billing year*	varies

State	Subscriber limit (% of peak)	Power limit Res/Com(kW)	Monthly rollover	Annual compensation
Pennsylvania	no limit	50/3,000 or 5,000	yes, until end of billing year	"price-to-compare" (generation and transmission cost)
Rhode Island	2	1,650 for most, 2250 or 3500*	optional	slightly less than retail rate
South Carolina	0.2	20/100	yes, until summer billing season	time-of-rate use or less
South Dakota	N/A	N/A	N/A	N/A
Tennessee	N/A	N/A	N/A	N/A
Texas***	no limit	20 or 25	no	varies
Utah	varies*	25/2,000 or 10*	varies—credits expire annually with the March billing*	avoided cost or retail rate*
Vermont	15	250	yes, accumulated up to 12 months, rolling	retail rate
Virginia	1	10/500	yes, avoided cost option at end of billing year	retail rate
Washington	0.5	100	yes, until end of billing year	retail rate
West Virginia	0.1	25	yes, up to 12 months	retail rate
Wisconsin	no limit	20	no	retail rate for renewables, avoided cost for non-renewables
Wyoming	no limit	25	yes, avoided cost at end of billing year	retail rate

Note: Some additional minor variations not listed in this table may apply.

N/A = Not available.

Lost = Excess electricity credit or credit not claimed is granted to utility.

Retail rate = Final sale price of electricity.

Avoided cost = "Wholesale" price of electricity (cost to the utility).

* Depending on utility.

** Massachusetts distinguishes policies for different "classes" of systems.

*** Only available to customers of Austin Energy, CPS Energy, or Green Mountain Energy (Green Mountain Energy is not a utility but a retail electric provider; according to www.powertochoose.com).

Source: Freeing the Grid

Resource Guide

Books

Brown, Lester, Emily Adams, Janet Larsen, and J. Matthew Roney. *The Great Transition: Shifting from Fossil Fuels to Solar and Wind Energy*. Wiley, 2014. I haven't read this book, but with Lester Brown at the helm, it ought to be a great read.

Bartmann, Dan and Dan Fink. *Homebrew Wind Power: A Hands-on Guide to Harnessing the Wind*. Buckville Publications, LLC, 2008. A clear and comprehensive guide to building a quiet, efficient, reliable, and affordable wind turbine.

Chiras, Dan. *The Homeowner's Guide to Renewable Energy*. Second edition. New Society Publishers, 2011. Contains information on residential wind energy and solar electric systems.

——. *Green Home Improvement: 65 Projects that Will Cut Utility Bills, Protect Your Health, and Help the Environment*. RS Means, 2008. Contains numerous simple projects to make your home more energy efficient and reduce your energy bills.

Ewing, Rex A. *Power with Nature: Solar and Wind Energy Demystified*. PixyJack Press, 2003. Skip to the second half if you want to get to the meat of the matter.

Gipe, Paul. *Wind Power: Renewable Energy for Home, Farm, and Business*. Chelsea Green, 2004. Somewhat technical introduction to small and large (commercial) wind generators.

——. *Wind Energy Basics: A Guide to Small and Micro Wind Systems*. Chelsea Green, 1999. A brief and somewhat technical introduction to wind energy for newcomers.

National Renewable Energy Lab. *Wind Energy Information Guide*. University Press of the Pacific, 2005. A brief introductory pamphlet on wind energy.

Piggott, Hugh. *Wind Power Workshop: Building Your Own Wind Turbine*. Center for Alternative Technology, 2011. A guide for those who want to build their own wind turbines and towers by Europe's small-wind expert.

Rohatgi, Janardan S. and Vaughn Nelson. *Wind Characteristics: An Analysis for the Generation of Wind Power*. Alternative Energy Institute, 1994. The best book available on siting wind turbines. Contains lots of math. You can obtain a copy from the American Wind Energy Association, or from the Alternative Energy Institute at (806) 651-2295.

Scheckel, Paul. *The Home Energy Diet*. New Society Publishers, 2005. A great guide for energy conservation in homes.

Shea, Kevin and Brian Clark Howard. *Build Your Own Small Wind Power System*. McGraw Hill, 2012. A book that is more about small wind systems you can buy than building your own. Full of lots of useful information.

Wegley, H.L., James V. Ramsdell, Montie M. Orgill, and Ron L. Drake. *A Siting Handbook for Small Wind Energy Conversion Systems*. WindBooks, 1980. Out-of-print guide to siting a wind turbine.

Woofenden, Ian. *Wind Power for Dummies*. For Dummies, 2009. Fact-filled book on small wind systems by one of the industry's leading authorities.

Articles

Brown, Lester R. "Harnessing the Wind for Energy Independence," *Solar Today* 16 (2): 2002, 24–27. Good overview of the prospects for commercial wind energy.

Dankoff, Windy. "Lightning Happens," *Home Power* 107, 2005, 60–64. Important information on protecting a renewable energy system from lightning.

Dankoff, Windy. "Top Ten Battery Blunders," *Home Power* 115, 2006, 54–60. Important reading for anyone installing a battery bank.

Davidson, Kelly. "Enervee: Scoring High Marks for Energy Efficiency," *Home Power* 161, 2014, 14. An interesting look at a new service that will help us determine the efficiency of appliances and electronic devices.

Gocze, Tom. "Heat-Pump Water Heaters," *Home Power* 156, 2013, 58–64. Another must-read for anyone who wants a much less costly and more efficient alternative for solar hot water.

Green, Jim. "Small Wind Installations in Colorado," *Solar Today* 14 (1): 2000, 20–23. Several case studies of wind installations.

Green, Jim and Mick Sagrillo. "Zoning for Distributed Wind Power—Breaking Down the Barriers." Conference Paper NREL/CP-500-38167, May 2005. National Renewable Energy Association, Golden, Colorado.

Gipe, Paul. "Small Wind Systems Boom," *Solar Today* 1 (2): 2002, 29–31. Good overview of small wind energy systems by one of America's leading experts on the subject.

Hackleman, M. and C. Anderson. "Harvest the Wind," *Mother Earth News* 192, 2002, 70–78. An easy-to-read and fact-filled article on wind power.

Home Power Staff. "Clearing the Air: Home Power Dispels the Top RE Myths," *Home Power* 100, 2004, 32–38. Superb piece for those who want to learn about the many false assumptions and beliefs regarding renewable energy.

Hren, Rebekah. "Monitoring Batteryless Systems," *Home Power* 164, 2015, 30–36. Overview of ways to monitor PV systems.

Kuebeck, Sr., Peter and Peter Kuebeck, Jr. "Old Jacobs—Current Again," *Home Power* 89, 2002, 70–78. Personal story about a great old wind turbine that's being refurbished and installed throughout the country.

Laughlin, Don. "You Gotta Have Height: A Tower Construction Project in Iowa," *Home Power* 92, 2003, 30–38. Interesting case study. Well worth reading to learn more about tower construction and erection.

Meyer, John and Joe Schwartz. "Battery Box Basics," *Home Power* 119, 2007, 50–55. Superb reference for individuals who want to build their own battery boxes.

Moore, Kevin. "Pumping Water with the Wind," *Home Power* 122, 2008, 88–93. A good overview of water-pumping windmills.

National Wind Coordinating Collaborative. "Wind Turbine Interactions with Wildlife and their Habitats," online publication, nationalwind.org. A very detailed look at studies on bird and bat mortality that helps set the record straight about large wind turbines.

Osborn, D. "Winds of Change." *Solar Today* 17 (6), 2003, 22–25. Looks at a number of important issues, includes bird mortality by wind towers compared to other sources.

Pahl, Greg. "Choosing a Backup Generator," *Mother Earth News* 202, 2004, 38–43. A fairly detailed overview of what to look for when buying a backup electrical generator for a RE system.

Pearen, Craig. "Brushless Alternators," *Home Power* 97, 2003, 68–71. Brief, but interesting, look at low-maintenance brushless alternators for wind machines.

Perez, Richard. "Flooded Lead-Acid Battery Maintenance," *Home Power* 98, 2004, 76–79. Read, memorize this, and put its advice into practice if you plan on installing a stand-alone RE system.

Piggott, Hugh. "Estimating Wind Energy," *Home Power* 102, 2004, 42–44. A brief piece that will help you determine how much electrical energy you can produce on your site.

Preus, Robert. "Thoughts on VAWTs: Vertical Axis Wind Generator Perspectives," *Home Power* 104, 98–100, 2005. Excellent article with excellent photos and diagrams.

Russell, Scott. "Starting Smart: Calculating Your Energy Appetite," *Home Power* 102, 2004, 70–74. Great introduction to using household load analysis to determine your household electrical demand.

Sagrillo, Mick. "Aesthetic Issues and Residential Wind Turbines," *Windletter* 23 (5): 2005, 1–2. A must-read for anyone who is interested in installing a wind turbine on his or her property. Many more of Mick's articles on small wind from *Windletter* are available at awea.org/smallwind/sagrillo

——. "An Open Letter: To Inventors of Vertical Axis Wind Turbines and Rooftop Wind 'Technological Breakthroughs,'" *Windletter* 27 (3): 2008, 1–6. A frank discussion about vertical axis wind turbines.

——. "Annual Wind Turbine Inspections," *Windletter* 21 (5): 2002, 1–3. Good advice for anyone who installs a wind turbine.

——. "Bats and Wind Turbines," *Windletter* 22 (2): 2003, 1–2. Important reading for anyone concerned about this issue.

——. "Buying Used Wind Equipment," *Windletter* 21 (11): 2002, 1–3. Sound advice on buying used wind turbines.

——. "Incremental Tower Costs versus Incremental Energy," *Windletter* 25 (1): 2006, 1–3. Looks at the cost of taller towers and the additional return on this investment.

———. "Payback: The Wrong Question," *Windletter* 26 (7): 2007, 1–3. Examines the topic of payback, especially why it is the wrong way to think about a renewable energy system.

———. "Planning Your Wind System (2)—Building Permits," *Windletter* 25 (7): 2006, 1–4. Will help you understand building permits and what you'll need to consider when applying for a permit.

———. "Planning Your Wind System—Evaluating Your Wind Resource," *Windletter* 26 (1): 2007, 1–3. Examines issues such as load assessment, wind speed monitoring, and wind maps.

———. "Planning Your Wind System—Homeowner's Insurance," *Windletter* 25 (11): 2006, 1–3. What every homeowner interested in installing a wind system needs to know about homeowner insurance to cover cost of damage to a wind turbine.

———. "Planning Your Wind System—Liability Insurance," *Windletter* 25 (10): 2006, 1–2. What every homeowner interested in installing a wind system needs to know about liability insurance.

———. "Planning Your Wind System—Utility Requirements," *Windletter* 25 (8): 2006, 1–3. Sound advice on working with the utility when installing a grid-connected wind system.

———. "Planning Your Wind System—Working with Your Neighbors," *Windletter* 27 (1): 2008, 1–4. Read this before announcing to your neighbors that you're going to install a wind turbine—in fact, before you buy your turbine and tower.

———. "Residential Wind Turbines and Lightning," *Windletter* 22 (11): 2003, 1–3. Describes lightning protection measures for small wind turbines.

———. "Residential Wind Turbines and Noise," *Windletter* 23 (3): 2004, 1–4. Important information on sound levels and allaying concerns of neighbors.

———. "Rules of Thumb for Siting Wind Turbines," *Windletter* 25 (6): 2006, 1–3. Good description of things you need to know when determining tower height for a wind energy system.

———. "Tall Tower Economics," *Windletter* 25 (2): 2006, 1–4. Discusses the incremental increase in energy output of turbines compared to the incremental cost of installing taller towers.

———. "The Wind's Power at Various Tower Heights," *Windletter* 24 (12): 2005, 1–3. A great discussion of tower heights that helps beginners realize the importance of tower height when installing a wind machine.

———. "Wind Matters," *Home Power* 158, 2014, 62–68. A very important article on the physics and math of wind that inform intelligent decisions when it comes to siting a wind turbine.

Sagrillo, Mick and Ian Woofenden. "Wind Turbine Buyer's Guide," *Home Power* 119, 2007, 34–40. A brief overview of wind turbines and details on several popular wind turbines.

Schwartz, Joe. "Finding the Phantoms," *Home Power* 117, 2007, 64–67. A must-read article for anyone interested in slashing electrical energy use, especially folks interested in going off grid.

———. "What's Going On—The Grid?" *Home Power* 106, 2005, 26–32. Excellent look at many of the grid-tied inverters.

Short, Walter and Nate Blair. "The Long-Term Potential of Wind Power in the U.S." *Solar Today* 17 (6), 2003, 28–29. Important study of wind power's potential.

Sindelar, Allan and Phil Campbell-Graves. "How to Finance Your Renewable Energy Home," *Home Power* 103, 2004, 94–99. Very useful article.

Stone, Lauri. "Hiring a PV Pro," *Home Power* 114, 2006, 48–51. Although this article is about hiring a professional installer for a PV system, much of what the author says applies to wind system installers.

Swezey, Blair and Lori Bird. "Buying Green Power—You Really Can Make a Difference," *Solar Today* 17 (1): 2003, 28–31. An in-depth look at ways (such as green tags) home-owners can tap into renewable energy without installing a system on their home.

Tobe, Jeff. "Managing Battery Charge Using Diversion Loads," *Home Power* 166, 2015, 52–58. Worthwhile reading for anyone interested in installing a small wind system that requires dump loads.

Weis, Carol and Christopher Freitas. "Battery System Maintenance and Repair," *Home Power* 161, 2014, 52–58. A must-read for anyone interested in a battery-based system.

Woofenden, Ian. "Battery Filling Systems of the Americas: Single-Point Watering System," *Home Power* 100, 2004, 82–84. This article is a must for those who would like to reduce battery maintenance.

———. "Managing Energy Use, One Load at a Time," *Home Power* 165, 2015, 16 and 18. An look at a very important part of renewable energy, load management.

———. "Off or On Grid? Getting Real," *Home Power* 128, 2008, 40–45. An in-depth look at the pros and cons of grid-connected and off-grid renewable energy systems. Important reading for anyone trying to decide whether to go off grid.

———. "Wind Generator Tower Basics," *Home Power* 105, 2005, 64–68. Excellent overview of tower types.

Woofenden, Ian and Mick Sagrillo. "How to Buy a Wind-Electric System," *Home Power* 122, 2007, 28–34. Terrific review of the steps you need to take when contemplating a wind turbine, plus a great overview of 25 residential wind turbines.

Woodenden, Ian and Roy Butler. "Wind Turbine Buyer's Guide," *Home Power* 161, 2014, 34–42. This article focuses on the companies selling wind turbines and is well worth reading.

Woodenden, Ian and Roy Butler. "Wind Turbine Buyer's Guide," *Home Power* 167, 2015, 50–58. Read this. It's worth its weight in gold.

Magazines

Backwoods Home Magazine P.O. Box 712, Gold Beach, OR 97444. Tel: (800) 835-2418. Website: backwoodshome.com. Publishes articles on all aspects of self-reliant living, including renewable energy strategies.

The CADDET Renewable Energy Newsletter 168 Harwell, Oxfordshire OX11 ORA, United Kingdom. Tel: +44 123335 432968. Website: cadet-re.org. Quarterly magazine published by the CADDET Centre for Renewable Energy. Covers a wide range of renewable energy topics.

EERE Network News Newsletter of the Department of Energy's Energy-Efficiency and Renewable Energy Network. You can read current and past issues on their website: apps1.eere.energy.gov/news

Home Energy Magazine 1250 Addison Street, Suite 211B, Berkeley, CA 94702. Be sure to check out their website: homeenergy.org. Great resource for those who want to learn more about ways to save energy in conventional home construction.

Home Power P.O. Box 520, Ashland, OR 97520. Tel: (800) 707-6585. Website: homepower .com. Publishes numerous extremely valuable how-to and general articles on renewable energy, including solar hot water, PVs, wind energy, micro hydroelectric, and occasionally an article or two on passive solar heating and cooling. This magazine is a goldmine of information, an absolute must for anyone interested in learning more. It also contains important product reviews and ads for companies and professional installers. They also sell CDs containing back issues.

Mother Earth News 1503 SW 42nd St., Topeka, KS 66609. Website: motherearthnews. com. One of my favorite magazines, but they are not publishing many articles on renewable energy any more.

Solar Today ASES, American Solar Energy Society, 2525 Arapahoe Ave, Ste E4-253, Boulder, CO 80302. Tel: (303) 443-3130. Website: solartoday.org. This magazine, published by the American Solar Energy Society, contains lots of good information on solar and a few articles on wind energy. Mick Sagrillo writes a monthly column for them on small wind energy. They also have some information on small wind energy on their website, too, which is worth reading.

Organizations

American Wind Energy Association. 1501 M St. NW, Suite 1000, Washington, DC 20005. Tel: (202) 383-2500. Website: awea.org. This organization sponsors an annual conference on wind energy. Check out their website which contains a ton of useful information on wind energy.

British Wind Energy Association. 26 Spring Street, London W2 1JA. Tel: 0171 402 7102. Website: britishwindenergy.co.uk. Actively promotes wind energy in Great Britain. Check out their website for fact sheets, answers to frequently asked questions, links, and a directory of companies.

Center for Alternative Technology. Machynlleth, Powys SY20 9Az. Tel: 01654 703409. Website: cat.org.uk. This educational group in the United Kingdom offers workshops on renewable energy, including wind, solar, and micro hydro.

Energy Efficient Building Association. 9900 13th Avenue N., Suite 200, Plymouth, MN 55441. Tel: (952) 881-1098. Website: eeba.org. As its name states, this organization focuses on building energy-efficient buildings. It offers excellent conferences, workshops, publications, and an online bookstore.

European Wind Energy Association. Rue d'Arlon 80, B-1040 Brussels, Belgium.

Tel: +32 2 213 1811. Website: ewea.org. Promotes wind energy in Europe. Their website contains information on wind energy in Europe and offers a list of publications and links to other sites.

National Wind Technology Center of The National Renewable Energy Laboratory. 1617 Cole Blvd., Golden, CO 80401-3393. Tel: (303) 275-3000. Website: nrel.gov/wind. Their website provides a great deal of information on wind energy, including a wind resource database.

Small Wind Certification Council. 500 New Jersey Avenue, NW, 6th Floor, Washington, DC 20001. Tel: (888) 422-7233. Website: smallwindcertification.org, info@smallwindcertification.org. Certifies small wind turbines in an effort to provide more useful information on wind turbines.

Solar Energy International. 39845 Mathews Lane, Paonia, CO 81428. Tel: (970) 527-7657. Website: solarenergy.org. Offers a wide range of workshops on solar energy.

Solar Living Institute. Solar Living Center, 13771 S Highway 101, Hopland, CA 95449. Tel: (707) 744-2017. Website: solarliving.org. A nonprofit organization that offers frequent hands-on workshops on solar energy and other topics. Be sure to tour their facility if you are in the neighborhood.

US Department of Energy and Environmental Protection Agency's ENERGY STAR program. Website: energystar.gov. This website contains invaluable information on Energy Star rated products. Start here when shopping for new electronics, like TV sets, as well as appliances, heating and cooling equipment, office equipment, lights, fans, and so on.

US Department of Energy's Energy Efficiency and Renewable Energy website: eere. energy.gov. Be sure to check out this website for information on energy efficiency, renewable energy, and transportation. This is a great spot to keep up to date on new developments.

Small Wind Turbine Manufacturers

Bergey WindPower, bergey.com
Bornay Wind Turbines, bornay.com
Britwind (British Wind), britwind.co.uk
Ecocycle, Ecocycle.com
Luninous Renewable, luminousrenewable.com
Pika Energy, Pika-energy.com
Kestrel Wind Turbines, kestrelwind.co.za
Marlec Engineering Co. Ltd., marlec.co.uk
Osiris, osirisenergy.com
Ventera, venterawind.com
Wind Turbine Industries Corp., windturbine.net
Weaver Wind Energy, weaverwindenergy.com
Xzeres, xzeres.com

Notes

Chapter 1

1. Ice buildup increases the drag in relation to the aerodynamic lift, which dramatically slows the blades down.

Chapter 2

1. Humid air is less dense because water (H_2O) has a molecular weight of 18. The main components of air, nitrogen (N_2) and oxygen (O_2), have molecular weights of 28 and 32, respectively. So, as water vapor displaces nitrogen and oxygen molecules, the air gets less dense.
2. Manufacturers routinely report rotor diameter. The rotor diameter is the diameter of the circle described by the spinning blades and is often greater than twice the "blade length." That's because the blades attach to the hub.

Chapter 3

1. In free-standing towers, these functions are provided by the tower itself and its foundation, so both the tower and the foundation must be much stronger.
2. Technically, the term grid refers to the high-voltage transmission system, not local electrical distribution systems; customers are connected to the grid through their local utility and indirectly to its distribution system, the grid. So, the terms *grid-connected* and *grid-tied*, which are used by the renewable energy community, are not entirely accurate. *Utility-connected* or *utility-tied* would be better terms, but that is not the common usage in the renewable energy world.
3. Although most small wind turbines produce wild AC, commercial-size wind turbines and one small wind turbine, the 5-kW Endurance, are *induction* turbines; they produce grid-compatible AC electricity.

Chapter 4

1. State wind maps contain estimates, not direct measurements of wind speed. Direct measurements are used to validate the maps, but maps are created using upper air wind speed data (wind balloon data) and topographic data. Computer models then predict the wind speed at 50 meters for the entire state.

2. As long as both height numbers are in consistent units (meters or feet) and both speed numbers are in consistent units (meters per second or miles per hour), the results will be OK, even if height is in feet and speed is in meters per second.

Chapter 5

1. Some vertical furling wind turbines like the Aeromax Lakota have a spring that complements gravity return.
2. In airplanes with variable-pitch propellers, the engine and propeller are operated at a relatively constant speed, and power is controlled by varying the pitch of the propellers.
3. An axial flux alternator consists of a round plate that contains the permanent magnets. It is attached perpendicularly to the rotor.

Chapter 10

1. Some states, like Wisconsin, have state laws that supersede local zoning authority, allowing wind turbines to be installed on tall towers for optimum performance.

Index

About the Author

DAN CHIRAS received his Ph.D. in reproductive physiology from the University of Kansas School of Medicine in 1976. Since then, he has taught at numerous colleges and universities, including The University of Colorado in Denver, the University of Denver, the University of Colorado in Boulder, the University of Washington, and Colorado College. He currently teaches part time through the Sustainable Living Department at Maharishi University in Fairfield, Iowa. Dan teaches courses on ecological design and renewable energy.

Dan is founder and lead instructor at The Evergreen Institute's Center for Renewable Energy and Green Building in Gerald, MO (evergreeninstitute.org). He teaches courses on solar electricity, wind energy, passive solar heating and cooling, solar hot water, natural building, energy conservation, natural plasters, green building, and net zero energy building.

Dan has published close to 400 blogs and articles in a variety of publications including *Home Power*, *Solar Today*, and *Mother Earth News*. He has also published 36 books including *Power from the Sun*, *The Solar House*, *The Homeowner's Guide to Renewable Energy*, and *The Natural House*. Dan has also published several college textbooks, including *Environmental Science*, *Natural Resource Conservation*, *Human Body Systems*, and *Human Biology*.

Dan's four newest books are *High-Performance, Off-Grid Chinese Greenhouses*, *Things I Learned too Late in Living Comfortably Off Grid*, and *Survive in Style: The Prepper's Guide to Living Comfortably Through Disasters*. Dan also recently co-produced a DVD on off-grid aquaponics in conjunction with The Aquaponics Source.

Dan and his partner Linda raise grass-fed beef, free-range chickens and ducks, and organic vegetables which they market locally. Their home, The Evergreen Institute, and the farm are all powered 100% by solar electricity, wind energy, and solar hot water. In 2013, Dan built a passive-solar, net zero energy home on the farm to replace their home that had been completely destroyed in a fire.

Dan is an avid reader and musician. He plays several instruments including the guitar, flute, saxophone, ukulele, and six-string banjo. Dan has written numerous songs that he is currently trying to publish.

A Note About the Publisher

NEW SOCIETY PUBLISHERS (www.newsociety.com), is an activist, solutions-oriented publisher focused on publishing books for a world of change. Our books offer tips, tools, and insights from leading experts in sustainable building, homesteading, climate change, environment, conscientious commerce, renewable energy, and more—positive solutions for troubled times.

Sustainable Practices for Strong, Resilient Communities

We print all of our books and catalogues on **100% post-consumer recycled paper**, processed chlorine-free, and printed with vegetable-based, low-VOC inks. These practices are measured through an Environmental Benefits statement (see below). We are committed to printing all of our books and catalogues in North America, not overseas. We also work to reduce our carbon footprint, and purchase carbon offsets based on an annual audit to ensure carbon neutrality.

Employee Trust and a Certified B Corp

In addition to an innovative employee shareholder agreement, we have also achieved B Corporation certification. We care deeply about *what* we publish—our overall list continues to be widely admired and respected for its timeliness and quality—but also about *how* we do business.

For further information, or to browse our full list of books and purchase securely, visit our website at: **www.newsociety.com**

New Society Publishers

ENVIRONMENTAL BENEFITS STATEMENT

For every 5,000 books printed, New Society saves the following resources:[1]

46	Trees
4,168	Pounds of Solid Waste
4,585	Gallons of Water
5,981	Kilowatt Hours of Electricity
7,576	Pounds of Greenhouse Gases
33	Pounds of HAPs, VOCs, and AOX Combined
12	Cubic Yards of Landfill Space

[1] Environmental benefits are calculated based on research done by the Environmental Defense Fund and other members of the Paper Task Force who study the environmental impacts of the paper industry.

 Certified B Corporation

 FSC www.fsc.org — MIX Paper from responsible sources FSC® C016245

 new society PUBLISHERS www.newsociety.com